ORIGIN OF EVERYDAY THINGS

JOHNNY ACTON,
TANIA ADAMS AND MATT PACKER

THINK
BOOKS

First published in Great Britain by Think Books in 2006.
This edition published in 2007 by Think Books
an imprint of Pan Macmillan Ltd
Pan Macmillan, 20 New Wharf Road, London N1 9RR
Basingstoke and Oxford
Associated companies throughout the world
www.panmacmillan.com
www.think-books.com

ISBN: 978-1-84525-048-5

Author: Johnny Acton, Tania Adams and Matt Packer
Editor: Jo Swinnerton
Think Books: Victoria Chow, James Collins, Rica Dearman, Serena Humphryes, Emma Jones, Lou Millward, Jackie Scully, Mark Searle, Rob Turner and Fay Warrilow

1 3 5 7 9 8 6 4 2

A CIP catalogue record for this book is available from the British Library.

Printed in the UK by Mackays of Chatham

Cover images: Comstock Images / Alamy, Jupiterimages
Other Images: Jupiterimages, United States Patent and Trademark Office, Mary Evans Picture Library, The Advertising Archives, The H J Heinz Company Ltd Collection held at the History of Advertising Trust Archive (www.hatads.org.uk), History of Advertising Trust Archive (www.hatads.org.uk), oldcomputers.net, Jon Santa Cruz/Rex Features, SNAP/Rex Features

'You can't sit on the lid of progress. If you do, you will be blown to pieces.'

Henry Kaiser, American industrialist (1882-1967)

THANKS

The authors would particularly like to thank IBM for developing the word processor and Tim Berners-Lee for inventing the World Wide Web.

Special thanks are extended to the hardworking team of Think Interns who added their own Origin-al touches to this book:

Daniele Bora, Paul Cannon, Elizabeth Chase, Tristan Cox, Kaylee Hultgren, Sonia Juttla, David Kircaldy, Chandni Lakhani, Lucy Moakes, Sam Riviere, Shereen Sally and Nuria Stylianou.

INTRODUCTION

As the saying goes, every object tells a story. Sometimes the best tales are told by the things that we most take for granted. Who would have thought, for example, that when the woman who brought the table fork to Europe died prematurely, the consensus was that it served her right for having such pretentious eating habits? Or that lipstick is 70,000 years old? Or that the first escalator, which raised its passengers all of seven feet, was considered one of Coney Island's most daring rides?

The history of innovation is full of such surprises. Some inventions are much more ancient than might be supposed, like the shopping mall, which made its debut in the second century BC, and the fax, which was first sent in 1843. Some everyday items, such as catseyes and balloons, have their origins in moments of breathtaking inspiration; others came into existence by sheer fluke. The catalyst for the microwave oven, for instance, was a chocolate bar melting in an engineer's pocket when he stood in front of a piece of radar equipment. The stories of everyday things also involve some unlikely celebrities. The man responsible for the cardigan, for example, was the soldier who ordered the disastrous Charge of the Light Brigade, and Liquid Paper was invented by the mother of one of The Monkees.

This book celebrates one of the defining features of our species: our ingenious ability to manipulate the physical word. It resolves burning issues such as who invented sliced bread and provides answers to questions you had forgotten you needed to ask. Above all, in revealing the startling life behind 400 of the items with which we are surrounded, *Origin of Everyday Things* proves that the things we might be tempted to view as mundane are usually anything but.

Johnny Acton

A

AEROSOL CAN

The French were the first to come up with the idea of combining liquids and gases for commercial purposes when confectioners began to carbonate sweet drinks with CO_2 in 1790. Then in 1837 Parisian Antoine Perpigna was responsible for the first soda siphon with a valve called the *Vase Syphoide*, which became a popular means of dispersing bottled water. By 1899 German boffins Helbling and Pertsch had patented the aerosol can, using methyl and ethyl chloride as propellants.

In 1926, Erik Rotheim (1898-1938) from Oslo, Norway, invented the first aerosol that could hold and dispense a variety of products. It was also the first refillable container to have a valve and propellant system, and he later remodelled his design to include a nozzle.

Rotheim quickly sold his idea to an American firm for 100,000 kroners, but it wasn't until 1941 that the duo Lyle Goodhue and William Sullivan deployed it to good effect in World War II among US troops in the South Pacific, enabling them to repel swarms of malaria-infested bugs. Between 1945 and 1950, the disposable aerosol went on the market and Edward Seymour began to manufacture spray-paint cans following a suggestion from his wife.

A soda siphon, one of the precursors to modern aerosol cans.

Sadly, the invention had a serious downside. In the mid-1970s, scientific research began to indicate that chlorofluorocarbons, or CFCs – gases that had become industry-standard as aerosol propellants – were adversely affecting the planet's protective ozone layer. By 1978 most American aerosol products were banned, and on 1 January 1989, the

Montreal Protocol, an international treaty encouraging 46 signatory nations and 31 ratifiers to phase out their CFC use, came into effect.

AIRBAG

A patent for a rudimentary airbag design was issued to American naval engineer John W Hetrick in 1953, but the crash-sensing technology necessary to make the device practical was perfected by US inventor Allen K Breed (1927-1999) in 1968. The first car sold to the general public with an airbag as a standard attachment was the 1973 Oldsmobile Tornado, but it was available only on the passenger side. Motor giant General Motors introduced the option of driver's side airbags in Oldsmobiles in 1975 and in Buicks the following year. In the early days, the devices were something of a liability in themselves – several fatalities were found to have been caused by malfunctioning airbags – but the major design issues had been solved by the early 1990s. One significant breakthrough was a venting system designed to reduce the rigidity of inflated airbags co-patented by Allen Breed in 1991. In the US, all new cars have been required to be equipped with airbags since 1998.

AIR CONDITIONING

People living in hot climates have taken measures to cool their environments since ancient times. The Babylonians splashed water on the outside of their dwellings, knowing that as it evaporated, it would draw heat from the interiors. In blazing summer conditions, wealthy Egyptians would order slaves to smear water on the outside of earthenware receptacles until evaporation caused their contents to cool. Modern air conditioning, however, owes its existence to the British scientist Michael Faraday (1791-1867), who discovered in the early nineteenth century that air could be chilled by compressing ammonia, allowing it to evaporate, condensing it again and repeating the process.

The first person to make practical use of Faraday's discovery was an American physician named John Gorrie (1803-1855). In 1841, he developed a system to blow cool air over his patients that used tubs of ice suspended from the ceiling during a yellow fever outbreak in Apalachicola, Florida. However, no complete description of this method has survived. Despite obtaining a patent for a Faraday based system in 1851, Gorrie failed to secure financial backing and the idea became dormant. It was resurrected in 1902 by Willis Haviland Carrier (1876-1950), who built the world's first modern air conditioning machine to address a problem at a printing plant in Brooklyn, New York. In hot, humid weather, the paper used for colour printing would absorb moisture, which messed up the alignment. Carrier countered the difficulty by developing a machine that dehumidified the air inside the plant by means of a cold water spray. Constant humidity was achieved via a 'dew point control', which varied the temperature of the spray according to the conditions.

Michael Faraday, one of the fathers of electricity, whose work in repeatedly condensing ammonia to chill air led to the invention of modern air conditioning.

ALKA-SELTZER®

The invention of Alka-Seltzer was nicely timed to coincide with the end of Prohibition (1920-1933), although the remedy actually hit the shop shelves in 1931, two years before the repeal of the Eighteenth Amendment. Once Americans could openly acknowledge their hangovers, there was huge demand for a product with which to nurse them.

The Alka-Seltzer story began in Elkhart, Indiana in 1928. During a flu epidemic, AR 'Hub' Beardsley, the president of Miles Laboratories Inc, visited the offices of a local paper called the *Elkhart Truth* and was impressed to find that none of the employees had succumbed. On making inquiries, he learned that the editor had pre-empted the disease by making his staff drink a mixture of aspirin and bicarbonate of soda every day. When he returned to his office, Beardsley asked Maurice Treener, the head chemist at Miles Laboratories, to cook up a batch of tablets on similar lines. In 1929, Beardsley took his family on a Mediterranean cruise and brought a gross of the tablets with him. It was a rough trip, compounded by an outbreak of influenza, so Beardsley started handing out pills to his fellow passengers. All of them, with the exception of two who died from pneumonia, reported that the tablets had greatly ameliorated their seasickness. Beardsley knew he was onto a winner. The only problem was that the pills tended to explode in their containers. Once this hitch had been ironed out by the Miles technicians, Alka-Seltzer, as the product was christened, was ready for market. The tablets proved a great hit, not least because of the soothing 'plop, plop, fizz, fizz' noise they made when they were dropped into water.

ALUMINIUM FOIL

The moisture-proof metal foil used to wrap food and tobacco was originally made from an alloy of tin and lead, which is why many people still call the familiar household product 'tin foil'. However, for almost a century it has been made from aluminium. The changeover occurred in 1910 at an aluminum plant in the Swiss town of Kreuzlingen called Dr Lauber, Neher & Cie. It was owned by the aluminium manufacturers JG Neher & Sons, and the sons in question had worked with Dr Lauber to develop a continuous aluminium rolling process. This allowed them to produce large and extremely thin sheets of the metal, which were used as a packaging material.

Nine years later, a Virginian named Richard S Reynolds founded the US Foil Company in Louisville, Kentucky. Reynolds had acquired a thorough knowledge of foil technology while working for his uncle, the tobacco baron RJ Reynolds. When the price of aluminium began to fall towards the end of the 1920s, Reynolds was well placed to convert his factory to the use of the metal as a raw material. He was soon supplying aluminium foil to the tobacco and confectionary industries. In 1947, Reynolds Metals, as the company was now known, unveiled the first foil specifically designed for domestic purposes. It was a mere 0.0007in thick and was an instant hit.

ANIMATION

The forerunners of the modern cartoon include the flip book and the phenakistoscope. This device, invented by the Belgian physicist Joseph Plateau (1801-1883) in 1832, comprised two circular discs, one incised with slots and one decorated with drawings of successive action. When the discs were rotated and the second was viewed through the slots in the first, the figures on the former appeared to move. William G Horner's (1783-1837) zoetrope worked on similar principles. Shaped like a child's drawing of a crown, it consisted of a slotted cylinder with images painted on the inside of the upwardly projecting parts. When it was spun around and the interior was viewed through the slots, an illusion of movement was generated.

Animation proper depended on the invention of motion picture films. The first cartoon in the modern sense was made by J Stuart Blackton (1875-1941) in 1906 and entitled *Humorous Phases of Funny Faces*. In 1914, Winsor McCay (1867-1934) introduced the first identifiable cartoon character to cinema audiences in Chicago. Her name was Gertie the Dinosaur.

ANORAK/PARKA

Rather confusingly, the garment known as a parka in the US tends to be called an anorak in the UK. This wasn't always the case. During the Mod era the British concurred with Americans in reserving the description 'parka' for a knee-length coat with a fur lined hood, as anyone who has seen the film *Quadrophenia* will know. Since then, however, the distinction has been blurred in Britain as 'anorak' has evolved into a derogatory term in for nerdy types given to train spotting. To add to the confusion, 'anoraks' often wear what are technically parkas.

A look at the origins of the two types of coat should settle the issue once and for all. Anorak, or more properly *Annuraaq*, is an Inuit word meaning 'item of clothing'. It refers to a buttonless hooded garment pulled over the head that is usually waterproof. Inuit anoraks are typically made from materials such as seal skins, animal intestines or the skins of Chinook salmon. Parka, on the other hand, is the word for 'skin' on the Aleutian Islands west of Alaska. The locals use it to refer to fur-lined coats in line with American usage.

ANTIPERSPIRANT

The first trademarked antiperspirant, Mum, went on sale in the US in 1888. It was invented by an anonymous man from Philadelphia, Pennsylvania, who marketed the product through his nurse. The original Mum was a waxy, zinc-based paste that was applied, somewhat messily, to the underarm area. Today the company markets considerably more sophisticated antiperspirants, which were once famously advertised under the slogan 'A girl's best friend is her Mum'.

The next major development was the introduction of Everdry in 1902. It was the first antiperspirant to incorporate aluminium chloride, setting a trend still followed by the industry. Nobody knew exactly why zinc- and aluminium-based compounds inhibited sweating, and scientists are still hazy about the precise mechanism that makes antiperspirants effective. The most popular theory is that the active ingredients block the sweat ducts, leading the body to temporarily abandon its attempts to sweat via a biofeedback mechanism.

In the early days, perspiration was such a taboo subject that requesting an antiperspirant at a pharmacy was a sweat-inducing experience in itself. It was not until 1919 that a manufacturer, Odo-Ro-No, dared to acknowledge the existence of body odour in its advertising. Interestingly, the first generation of antiperspirant buyers was exclusively female. Only in the 1930s did the industry begin to target male customers.

ARTIFICIAL LIMB

The oldest literary reference to a prosthetic limb occurs in the *Rig Veda*, a collection of more than 1,000 Sanskrit hymns compiled during the second

US Patent 1,350,518 24 August 1920

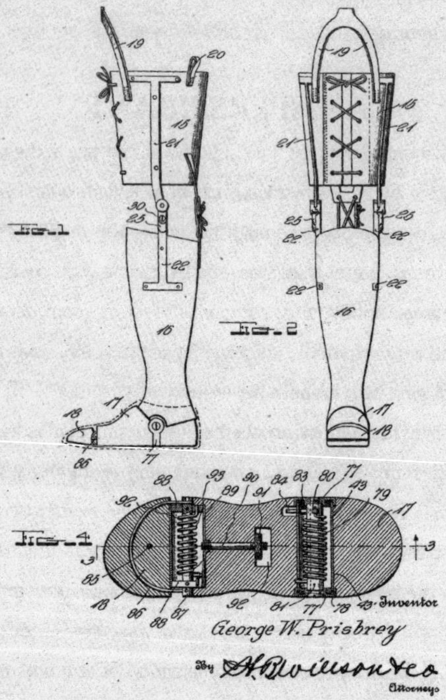

This early patent application for an artificial leg was awarded by the US Patent and Trademark Office in 1920

millennium BC. One of the hymns recounts the tale of a warrior queen called Vishpla who loses her leg in battle and replaces it with an iron substitute. Possibly more reliable is an account written by the Greek historian Herodotus (484-425BC) in about 440BC. It concerns a Persian seer named Hegistratus of Elis who was sentenced to death by the Spartans. Hegistratus supposedly escaped by cutting off his foot and slipping out of his leg iron. He is then said to have fashioned a wooden replacement and walked 30 miles before he was recaptured and beheaded.

The earliest actual example of an artificial limb was unearthed in Capua in Italy in 1858. It was a convincing looking leg made with a bronze covered wooden core and leather straps, and is believed to date from approximately 300BC. The prosthesis ended up in the British Museum but was destroyed in a bombing raid during World War II. Luckily, the museum had the foresight to make a copy in 1910.

ASPIRIN

Aspirin, or to give it its scientific name acetylsalicylic acid, is a man-made derivative of a substance called salicin which occurs naturally in the meadowsweet plant (*Spiraea ulmaria*) and the bark of willow trees. The great Greek physician Hippocrates (460-377BC) was prescribing powdered willow bark for pains, fevers and headaches as early as the fifth century BC. In 1838 an Italian chemist called Raffaele Piera isolated the active ingredient in salicin and gave it the name salicylic acid. Unfortunately it was extremely harsh on patients' stomachs. What was needed was a way of neutralising salicylic acid to make it fit for human consumption. The answer was provided in 1853 by the French chemist Charles Frederic Gerhardt (1816-1956), but at the time nobody thought of using his technique to develop a commercial product. It was resurrected in 1897 by a German named Felix Hoffman (1868-1946), who was desperately looking for something to alleviate his father's rheumatoid arthritis. The acetylsalicylic acid worked a treat. When Hoffman reported his discovery to his bosses at the Bayer drug company, they immediately decided to market the compound. Because the main source of the raw materials was the *Spiraea* plant, Bayer decided to call it 'Aspirin'. It was patented in 1899 before going on sale, initially as a loose powder, then in soluble tablet form.

ATLAS

The oldest surviving attempt at a map of the world is a Babylonian clay tablet in the British Museum dating from the sixth century BC. The five-inch map depicts Babylon as a rectangle at the centre of the world, bisected by the river Euphrates. The land mass is surrounded by an ocean, which has seven triangular islands at its outer periphery. Text inscribed on the tablet reveals that the Babylonians only had a hazy conception of the characteristics of these islands, but believed them to be inhabited by legendary beasts (one is home to a horned bull which 'attacks the newcomer') and to vary dramatically in brightness. On one of the islands, the light is said to be 'brighter than that of sunset or the stars', while another is a land where 'one sees nothing' and 'the sun is not visible'. These descriptions may reflect second-hand knowledge of conditions in the far north. Interestingly, the cartographer completely ignores Persia and Egypt, which were well known to the Ancient Babylonians.

The first printed work that can properly be described as an atlas was published in Bologna, Italy, in 1477. It was based on the calculations of Claudius Ptolemy (85-165AD), an Alexandrian geographer of the second century AD, and contained 27 maps. As the voyages of discovery soon made clear, however, there were significant omissions including Southern Africa, the East Indies, Japan and the Americas.

AUDIO CASSETTE

Thanks to the digital revolution, cassettes have largely become a thing of the past. But the special thing about cassettes, and the tape inside, was always their versatility – particularly blank cassettes, as Philips discovered in 1963 when it grossly underestimated consumer demand for them. As

US Patent 5,420,738 30 May 1995

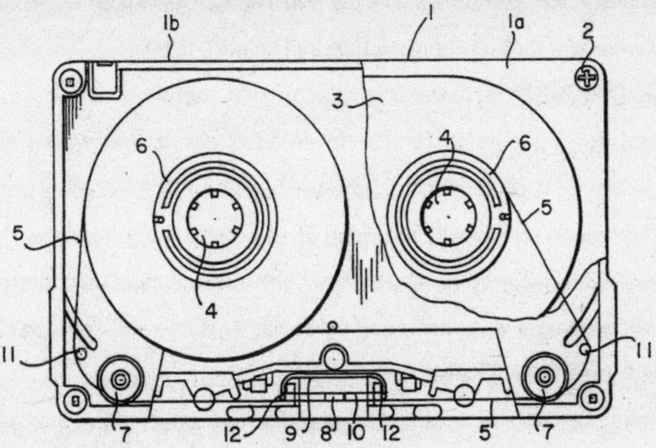

A cut-away plan view of the internal mechanism of an audio cassette, comprising a felt pad and a magnetic tape.

any child of the 1960s, 1970s or 1980s will remember, all those hours making mix-tapes in their bedrooms were as happy as they were wasted.

The first experiments with magnetic tape and recording were carried out in Germany, and High Fidelity (a clearer recording technique) was pioneered there during World War II. After the war, the Allied forces asked serviceman John Mullin (1913-1999) to investigate – read, purloin – German recording technology. At Radio Frankfurt, he discovered several 'magnetophones' and 1,000-metre reels of tape, which he mailed back to the US. 'I really flipped,' he recalled in an interview. 'I couldn't tell from the sound whether it was live or playback.'

In 1946, Mullin demonstrated the technology to members of the Istitute of Radio Engineers, and, as a result, received interest from Bing Crosby who wanted to use the tape to record. Tape allowed the artist to edit their recordings, and later to overdub tracks: technology that completely altered the sound of modern music. Though reel-to-reel became the industry standard, it was never a hit with consumers as the format was too bulky and impractical. What the listener needed was a tape that wasn't the size of a small briefcase. Philips duly shrank the technology, and compact cassettes that used tape only an eighth of an inch wide were in mass-production by the mid-1960s.

Ray Dolby (b1933) – the originator of that useful button on a whole generation of stereo systems – invented a circuit that raised music above tape's characteristic hiss, and the higher quality recording process was commercialised in the early 1970s. At around the same time, Henry Kloss (1929-2002) contributed chromium dioxide tape, and cassettes quickly became a vinyl-depleting epidemic. Portable and car tape decks helped spread the format.

While digital technology threatens to erase tape for good, increasingly, many recording artists are eschewing cold, hard digital technology for the warmth of tape; for example, the White Stripes' fourth album, *Elephant* (2003), was recorded entirely on analog equipment that was regarded as cutting edge in the mid-1970s. Many digital studios will record a finished song on to reel-to-reel and then back onto a computer for the 'enriching' effect only tape can provide. As with vinyl, which made a comeback in the hip-hop era, tape may be too good an invention to discard for ever.

AUTOMATIC FIRE ALARM/ SMOKE DETECTOR

The world's first automatic fire alarm was patented in 1902 by George Darby, an electrical engineer from Birmingham, England. Rather wonderfully, its most important component was butter. The device contained a pat of the familiar dairy product sandwiched between two metal plates. In the event of the temperature rising to danger level, the butter would melt, causing the top plate to fall onto the lower one. This would complete an electrical circuit, triggering a siren.

The battery-operated smoke alarm that now protects millions of homes worldwide was patented by Americans Kenneth House and Randolph Smith in 1969.

AUTOMATIC TELLER MACHINE

The world's first Automatic Teller Machine (ATM) was installed in 1969 at the Rockville Centre branch of Chemical Bank at 10 North Village Avenue in Long Island, New York. It was the brainchild of Don Wetzel,

US Patent 3,761,682 25 September 1973

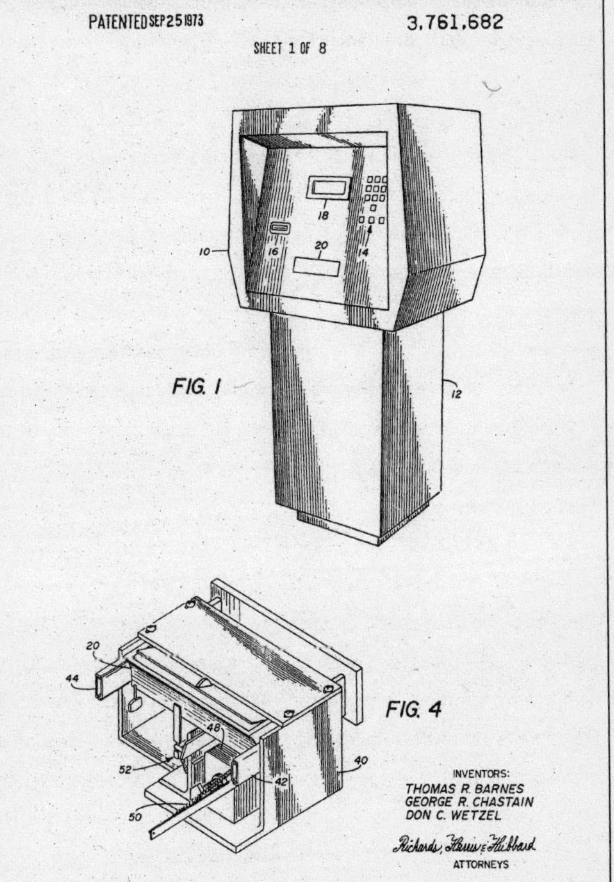

Don Wetzel's original ATM patent.

the former vice-president of product planning at Docutel, the company that developed automated baggage-handling equipment. Wetzel's moment of inspiration came while he was twiddling his thumbs in a long queue at his local bank.

A simpler version of the cash dispensing machine had been unveiled two years earlier at a branch of Barclays Bank in North London, but it worked by means of paper vouchers. The Chemical Bank's ATM was the first to use magnetic strip technology.

B

BABY FORMULA MILK

Long before Nestlé became famous for milk chocolate, the Swiss firm specialised in baby milk. The company's German-born founder Henri Nestlé (1814-1890) spent the 1860s searching for a substitute food for the infants of women who were unable to breastfeed. In 1867, he came up with *farine lactée*, a formula based in his own words on: 'wholesome Swiss milk and a cereal component baked by a special process of my invention.' Nestlé tested the product by feeding it to a dangerously ill premature baby. The little boy survived. Within five years, the inventor was exporting *farine lactée* as far afield as Australia and South America. Nestlé sold his company to Jules Monnerot in 1874 for one million francs.

BALLPOINT PEN

The ballpoint pen was originally designed in 1888 by American leather tanner, John Loud. His pen featured a reservoir of ink and a roller ball, and was used for marking up leather hides. However, it was never commercialised as the ink often clogged up and leaked. Laszlo Biro (1899-1995) created an improved version in 1935. Biro's design employed the same kind of ink used in newspaper presses to enable their pages to dry off quickly, leaving the paper dry and smudge-free. The pen was fitted with a tiny ball bearing that, when rotated, picked up the ink from a cartridge and left it on the paper. Biro sold the product in Argentina, where it was curiously unsuccessful. It was not until 1945 that the pen eventually took off. Milton Reynolds (1892-1976), from Chicago, copied Biro's design after his visit to Argentina and used Gimbels as the first store to sell the product.

The first ever ballpoint pen advert in *The New York Times* described the pen, somewhat ambitiously, as 'a miraculous fountain pen... guaranteed to write for two years without refilling'. On its release, Gimbels sold its stock of 10,000 ballpoint pens worth $12.50 each.

In England, the most common ballpoint pen is the Bic. Produced by Marcel Bich (1914-1994) In 1952, a French manufacturer of penholders and pen cases, the Bic ballpoint pen was inexpensive, non-leaky and

US Patent 6,305,865 23 October 2001

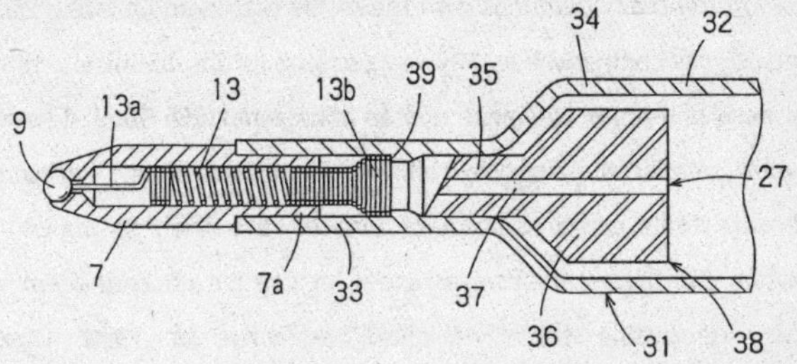

This sectional view of a retractable ballpoint pen shows the mechanism of a ballpoint pen refill with ink accommodating tube.

clear to read. More than 50 years later the Bic is still the most recognised ballpoint pen.

The latest style of ballpoint to reach the market is Yoropen. The award-winning design of this modern contender allows the writer to change the angle that the pen hits the paper. 'Not only is Yoropen stylish and funky,' says the pen manufacturer's Frankfurt-based director, Keith Cole, 'it can make writing more comfortable and pain-free for everyone.'

In the movies, particulary the *James Bond* kind, ballpoint pens have frequently contained hidden secrets, like explosives and radio transmitters. However, Wayne Leigh has produced the most comical secret use of a ballpoint pen to date. His 2003 idea, known as a 'Rep Pen', changes a ballpoint pen into a coat hanger, so you'll never crease your shirt – or forget your pen – again.

BAND-AID®

The Band-Aid was invented by a man with an accident-prone wife. In 1917 a cotton buyer named Earle Dickson (1892-1961), who worked for Johnson & Johnson in New Brunswick, New Jersey, married a lady called

Josephine Frances Knight. It quickly became apparent that she could scarcely enter the kitchen without burning or lacerating herself. At first, Dickson treated her wounds with the sterile cotton and gauze dressings made by his company, but they were so large (18 x 2.5 inches) that he felt he was using a sledgehammer to crack a nut. He decided to make several smaller dressings by affixing rectangles of the Johnson & Johnson sterile gauze to strips of surgical tape. Then, to prevent the plasters sticking to themselves ahead of use, he covered them with easily removable strips of crinoline fabric. In 1920, Dickson demonstrated the mini-dressings to his boss James Johnson (1856-1932). Johnson was impressed and promptly ordered production to commence. The name 'Band-Aid' was suggested by a company superintendent named W Johnson-Kenyon. At first, sales were sluggish, but then the marketing department had an inspired idea. They distributed free samples to Boy Scout troops throughout America, a ploy that secured the product's success. To date, more than 100 billion Band-Aids have been sold.

BANKING

In the ancient world, the safest places in which to deposit valuables were temples, which were well guarded and considered inviolable. The earliest references to activities that can be described as banking can be found in ancient Akkadian tablets from 2,500BC Babylonia, nine out of 10 of which are accounting records for deal like loans and mortgages. By the medieval era, moneylenders were a familiar sight in European cities. As a result of Christian disapproval of the practice of usury (lending money for interest), they were often Jewish, like Shylock in William Shakespeare's (1564-1616) *The Merchant of Venice*. Moneylenders typically clustered in town squares, where they worked from benches or tables. The term 'bank' derives from *banca*, the Italian word for bench. When a lender went bust, he would physically break his bench to indicate the demise of his business. This is the origin of the expression 'bankrupt', via the Italian *banca rotta* (broken bench).

Banking was once a far grander affair.

BARBIE DOLL™

In some circles, Barbie is regarded with deep suspicion. If she was blown up to adult proportions, her vital statistics would be in the order of 39-18-33 (a nigh on impossible measurement in human terms), and many believe her figure to be a pernicious influence on young girls because it sets an impossible ideal. Yet when she first appeared in 1959, Barbie represented something of a feminist breakthrough. For the first time in the modern era, girls had a doll to play with who could embody their adult aspirations, rather than just confirm their assumed future role as mothers.

In 1945, a young couple called Elliot and Ruth Handler (1916-2002) joined forces with their friend Harold Mattson to start a picture framing business in Hawthorne, California. Keen to branch out into toys and games, they started a new company and called it Mattel, from Mattson and Elliot. Significantly, the Handlers had two young children. Their names were Ken and Barbara.

As Ruth watched her daughter growing up during the 1950s, she noticed that Barbara and her friends seemed more interested in playing with adult dolls than with the traditional baby versions. Unfortunately, the only grown-up dolls available at the time were made of cardboard. Ruth became convinced that a three-dimensional alternative would be a commercial success. She took her idea to the board of Mattel, but it was rejected as being too risky and expensive. The executives only decided to back Handler's hunch when she returned from a trip to Switzerland with an adult doll called Lilli, who was based on a German cartoon character. Ruth's doll went into production and was named 'Barbie' in honour of her daughter. Barbie was unveiled at the 1959 American Toy Fair in New York. Mattel sold 351,000 Barbies in year one. Two years later, the company introduced Ken, a male counterpart named after the Handler's son. Several 'Barbie friends' followed, including Midge in 1962 and Skipper, Barbie's sister, in 1964. To date, more than a billion dolls in the Barbie range have been sold. If they were laid end to end, they would encircle the globe more than seven times.

BAR CODE

The bar code came into existence as a result of a conversation overheard in the halls of the Drexel Institute of Technology in Philadelphia. One day in 1948, a graduate student named Bernard Silver (1924-1962) listened with interest as the president of a chain of food stores begged one of the institute's deans to research a method of capturing product information at checkout counters. The professor declined, but Silver decided to take the matter further. He enlisted the help of a fellow graduate called Joseph Woodland (b1921) and began to work out a solution to the store president's dilemma. At first the pair experimented with a device that took readings by means of ultraviolet light, but they eventually devised a technique that

US Patent 2,612,994 October 7, 1952

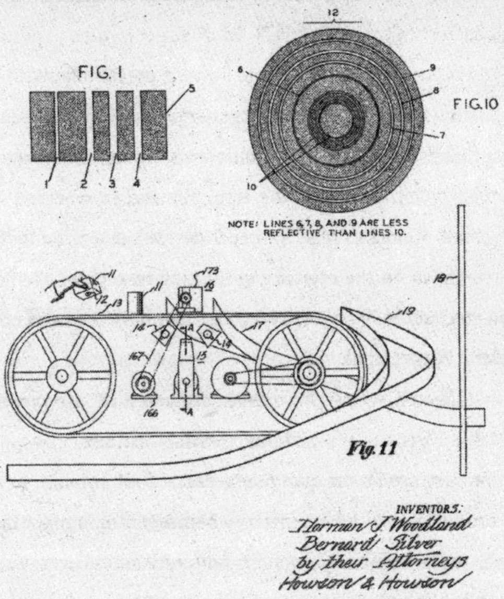

Woodland and Silver's original bar code system. Note the use of concentric circles (fig. 10) rather than parallel lines, allowing readings to be taken from any direction.

combined Morse code with a movie soundtrack system which was developed by Lee de Forest (1873-1961) during the 1920s. De Forest had printed patterns of varying opacity on the edges of film reels which were translated into sounds via light sensitive tubes as the movies ran.

Silver and Woodland filed their first barcode patent application on 20 October 1949. It outlined the so-called 'bull's eye' system in which the markings on products took the form of concentric circles rather than today's parallel lines. This allowed readings to be taken from any direction. Sadly, Silver died in 1963 at the age of 38, four years before barcodes were finally adopted by American retailers.

BAROMETER

The barometer was invented by an Italian mathematician named Evangelista Torricelli (1608-1647), who was a former pupil of Galileo (1564-1642). In 1643, Torricelli filled a long metal tube with mercury, placed a finger over the non-sealed end and stood the tube in a tray of the same liquid metal. He then withdrew his finger. Most of the mercury in the tube didn't flow into the tray, but stayed where it was in a column about 30 inches high. Torricelli decided that something had to be pushing down on the mercury in the tray to support the column in the tube. It could only be air, which had not hitherto been considered to have any weight at all.

The first person to use Torricelli's discovery to demonstrate that atmospheric pressure declines with altitude was another mathematician, the Frenchman Blaise Pascal (1623-1662). In 1648, he had his brother-in-law take a mercury barometer up a mountain in the Auvergne, while a friendly monk monitored a similar device in the brother-in-law's low-lying garden. The mercury column at the top of the mountain was three inches shorter than the one in the garden. Pascal had proved his point.

BATTERY

The spur to the invention of the battery was a debate about the existence or otherwise of a vital force called 'animal electricity'. The theory was propounded during the 1780s by an Italian anatomy professor called Luigi Galvani (1737-1798), who had produced contractions in the leg muscles of dead frogs by touching the nerves with strips of statically charged metal. The Italian physicist Alessandro Volta (1745-1827) was highly sceptical. In 1800, he built a device consisting of alternating disks of copper and zinc separated by pieces of brine-soaked cloth. When he attached a wire to either end of his creation, a continuous electrical current was generated. This proved that electricity was a force independent of living creatures. With minimal humility, Volta named the device the 'Voltaic Pile', but it later became known as a 'battery'. He also gave the first four letters of his surname to the well known electrical unit.

BICYCLE

The bicycle evolved into its familiar form by degrees. The first machine worthy of the name was perhaps the Draisienne or 'hobby horse' patented by the German Baron von Drais (1785-1851) in 1818. It looked not unlike a modern bike – there were two wheels of the same size mounted on a frame and the front one was attached to a handlebar for steering – but there were two crucial differences. The entire apparatus was made of wood and there weren't any pedals. The Draisienne was propelled forward by pushing the feet against the ground. This was all very well on paved paths in municipal parks, but not much use anywhere else.

A Scotsman called Kirkpatrick Macmillan (1812-1878) realised that if he could propel a 'hobby horse' along without putting his feet on the ground, what an improvement that would be. Macmillan's bicycle moved by a horizontal reciprocating movement of the rider's feet on the pedals.

The velocipede or "boneshaker."

Connecting rods helped transfer that energy to cranks on the rear wheels. Macmillan did not patent his technology, he was more content with living quietly and ignoring the furore his innovation stirred up. Things improved in 1863 with Pierre Lallement's (c1843-1891) invention of the velocipede or 'boneshaker'. This was the first cycle to feature pedals, added when it was mass-produced by Pierre Michaux (1813-1883) in 1867, which were directly affixed to the front wheel, but as the nickname implies the ride was far from comfortable.

The next innovations were metal spoked wheels and rubber tyres. These initially spawned a class of cycle with one enormous wheel and one tiny one. One such vehicle was the Penny Farthing, named after the largest and smallest British coins then in circulation. Safety bicycles, the first to feature chains, were popularised by a model called the Rover which hit the market in 1885. Three years later, John Dunlop (1840-1921) patented the pneumatic tyre, and by the 1890s, improvements in metallurgy permitted designers to once again equip bicycles with two wheels of equal size.

BIKER JACKET

The biker jacket, a symbol of youthful rebellion in the era of James Dean and Marlon Brando, began its life as an item of military dress. In World War I, the protective garb for motorcycle-riding personnel was a long coat known as a duster. This was a hard-wearing, waterproof coverall made of hide that would stand up to all kinds of battering from the elements. The duster coat also saw action on the American prairies, shielding the backs of cowboys from rough sand and wet weather. But in World War I, it proved to be flawed for the army riders: the coat's tail would often get caught up in the wheels of the bike, with predictable consequences. Something had to be done in order to make the coats more 'biker-friendly'.

Ross Langlitz (b1918) solved the problem. A motorcycle-crazy youth from Plymouth, Idaho, he lost a leg in a crash when he was just 17. Undeterred, he began to make biking leathers in his basement, and soon realised that he could make a living out of it. In 1947, Langlitz and his wife Mavis moved to Portland, Oregon and founded Langlitz Leathers. Its express purpose: to craft the finest bikerwear known to man.

Taking a germ of influence from the military coat – which, by then, had already been shortened – Langlitz customised it to withstand 75mph and thickened the leather so that it would protect a fallen rider. He also gave it some unusual features, which would later become iconic. Thick zippers crept onto the cuffs for added safety and security, the zipped pockets went diagonal, and the sleeves were lengthened to make room for the rider's reach. Langlitz called his jacket the 'Columbia', and sold it for $38.50. The next thing he knew, Marlon Brando was modelling it in *The Wild One* (1954), and its classic status was sewn up.

Never willing to sacrifice quality for quantity, Langlitz Leathers now makes an average of six garments per day. Though they are far from cheap, they are universally acclaimed by both bikers and fashion hounds, and the sheer number of imitators speaks volumes about the style's success.

BIKINI

When the bikini first appeared in the 1950s, it was met with a great deal of public outrage. But two-piece bathing costumes had been done before – a very long time before. Mosaics found in a villa from the early fourth century at Sicily's Piazza Armerina show images of young women dancing in what look like two-piece bathing suits, although it appears that the garments were simply two narrow pieces of cloth swaddled around the relevant attributes.

Sadly, this liberal attitude to female flesh gradually disappeared, at least in much of the Western world, and women were reduced to swimming in cumbersome outfits that did little to improve their buoyancy. It was not until after World War II that the two-piece made a comeback.

The idea of a two-piece bathing suit was hit upon by two Frenchmen at roughly the same time in 1946. First past the post was Jacques Heim (1889-1967), who sent a team of skywriters over the Cannes seafront

A 1950s bikini with an even more risqué strapless design.

to advertise his 'atome' – so called on account of its size – as 'The smallest bathing suit in the world.'

Rival designer and engineer Louis Reard (1897-1984) revealed his own design just three weeks later. He, too, employed a team of skywriters to announce his product over the French Riviera. Choosing the slogan, 'Smaller than the smallest bathing suit in the world', Reard also chose an exotic-sounding name for his outfit: Bikini.

Whether Reard had drawn this from his own reported claims that he had 'split the atome', or from the name of the South Pacific islands that were home to several US nuclear tests, is unclear – but it set off such a chain reaction of bomb-related headlines in the world's press that the UN must have longed to impose a 'Swimwear Puns Non-Proliferation Treaty'.

Reard may have been beaten to the punch, but he won the argument, coining a name that evoked all the sun-kissed locations in which the two-piece would flourish. Micheline Bernardini, his first model for the outfit, acquired 50,000 fan letters for her efforts. And with Brigitte Bardot's (b1934) scantily-clad turn in *And God Created Woman* (1958), and the release in 1960 of Brian Hyland's (b1943) song 'Itsy Bitsy Teenie Weenie Yellow Polka Dot Bikini', Reard's one-upmanship created a cultural icon.

BLOOD BANKS

The world's first blood bank was established in 1932 in a hospital in Leningrad (now St Petersburg) in Russia. The first US blood bank was opened a few years later, in the Cook County Hospital, Chicago in 1937. It was set up by the hospital's therapeutic director Bernard Fantus, who also invented the term 'blood bank'.

In 2005, the Chinese secured another blood bank 'first' when a unit solely dedicated to the giant panda was completed in Wolong Nature Reserve in the Szechuan province.

BOOTS

It was not until the mid-nineteenth century that mass production allowed poorer folks to partake of a fashion previously reserved for the rich. Fine boots had always been indicative of social status. Etruscan warriors went to battle barefoot, reserving the luxury of covered boots for soothsaying priests. The boots worn by Roman soldiers indicated military rank, depending on how far they went up the leg; Julius Caesar (c100-44BC) was said to have worn a pair fashioned out of gold. Throughout history, new styles of boots were coveted by royalty, and uncomfortable, often impractical designs were tolerated for the sake of looking fabulous.

Such was the case with the 'stivali', a popular lightweight boot that immigrated to England during the twelfth to fourteenth centuries. It was worn high and wide on the leg, often forcing the wearer to adopt a bowlegged gait. Not only did the boots hinder knee bending, but the excessive width at the top of the boot let the rain in. Another example occurred in sixteenth century France. Henry IV (1553-1610) was convinced that French craftsmen were incapable of satisfying his exquisite tastes in leather, so he commissioned a tanner to study the leatherwork of Hungary. His influence made boots highly fashionable in France, which led to their appearance in salons as well as dance floors. At times, these boots were also not so knee-friendly. To keep them as snug as possible they were first soaked in water. Once fitted and dried, the wearer often had to resort to waddling. It is possible there is a correlation between this look and the swaggering gait considered macho at the time.

An 1846 cartoon from *Punch*.

In the seventeenth century King William of Orange (1533-1584) introduced the jackboot – a sturdy contraption worn high above the knee. Though typically worn by soldiers and horsemen, thigh-high boots were also used by fishermen (before gum boots were around), and originally by pirates and smugglers. For once, someone wore a pair for practical purposes: the common practice of tucking booty inside their widened tops coined the term 'bootlegging'.

By the middle of the nineteenth century, boots became easier and cheaper to manufacture, and were no longer an indicator of social status. Men and women from all social and economic backgrounds could wear boots, whether for sensible or decorative reasons. This was not good news for boot and shoe craftsmen though. Fearing the bankruptcy of their trade, they became one of the most outspoken groups against the industrialisation of skilled craftsmanship. In an effort to combat the development of factories and their poorly paid (if paid at all) workers, competitions were staged to demonstrate the ineffectiveness of the factory worker. Some of the most refined work ever done was produced for these competitions; one pair made in Philadelphia boasted 64 stitches to an inch – twice as much as a modern sewing machine can manage.

In the early twentieth century, boots took on a new utilitarian image in the midst of the war and economic depression, and comfort and durability became desirable qualities among working men and women. Also around this time was the advent of the cowboy boot. Emerging as a distinct style in the 1880s, the original cowboy boot had no ornamentation and was worn very tight for maintaining control in the saddle. It wasn't until 1903 that the first embroidery appeared. Not surprisingly, plain boots couldn't hack it in Hollywood. Onscreen cowboys such as Tex Ritter (1905-1974) and Tom Mix (1880-1940) wore elaborate, invented styles (which were surely easier to walk in), based more on cheap novels and comic book illustrations than the actual Wild West. When real cowboys such as Wyatt Earp (1848-1929) appeared on the set of films like *The Half Breed* (1919), the fake styles paled in comparison.

BOWLING

Bowling may be one of the oldest sports in existence. The earliest evidence of the game was found in the 1930s by the British anthropologist Sir Flinders Petrie (1853-1942), who discovered a bowling-like game in a child's grave in ancient Egypt dating back to 3,200BC.

The first written reference to bowling dates back to 1366, when King Edward III (1312-1377) allegedly banned his troops from playing it so they were not distracted from their archery practice. It was certainly popular in the time of Henry VIII (1491-1547), who ordered lanes to be built in Whitehall. Many variations of 'pin' games evolved across Europe, such as Italian *bocce*, French *petanque*, and Britain's lawn bowling. One of the most eccentric and energetic forms is still found in Edinburgh. The player swings a ball down a lane and in doing so flops onto the lane on their stomach.

Medieval bowling games were apparently a lot more violent.

Today, its biggest cult following is in America. Quite how it caught on in the US is unknown, although presumably various settlers took their versions with them across the Atlantic. The earliest recorded mention of it in American literature is by Washington Irving (1783-1859), when Rip van Winkle awakens to the noise of 'crashing ninepins'. By the 1800s, bowling was catching on all over America. It was not until 1895 that restaurateur Joe Thum formed the American Bowling Congress, enabling standardisation and national competitions. In the early twentieth century, bowling technology improved dramatically. Up until this time, balls had been made of hard wood, but in 1905 the first rubber ball was produced. In 1952, the first automatic pin setting machine was produced, doing away with the need for 'pinboys'. Perhaps the biggest boost to bowling's popularity was NBC's *Championship Bowling* in the 1950s. It is now fully engrained in US culture and is one of America's most popular pastimes.

BRA

The need for uplift and support for the female form originates as far back as Ancient Greece. From as early as 2,500BC, the Minoan women of Crete wore garments that not only supported their breasts, but actually lifted them out of their dress, as was the fashion at the time. In comparison, the female athletes of Sparta relied on a 'mastodeton' or 'apodesmos', a band of cloth that strapped the breasts down, thus allowing ease of movement in sports. So it seems the Ancient Greeks not only invented the Olympic Games but the sports bra too.

In 1889, Herminie Cadolle in France invented the first modern bra. The garment was originally a two-piece called *le bien-être* or 'the well-being', consisting of a sturdy corset for the waist and shoulder straps to support the breasts. By 1905 the upper half was being sold separately without the corset.

There is some controversy over who invented the bra as we know it today. New York socialite Mary Phelps Jacob (1891-1970) developed the first actual patented bra in 1910. After purchasing a sheer evening gown

US Patent 24, 033 May 17, 1859

A very early (and uncomfortable-looking) predecessor to the modern bra.

for a big social event, Mary discovered that her usual whalebone corset simply didn't do the dress justice. The dress was too sheer for such a heavy undergarment, and so she made a makeshift bra using two silk scarves and some ribbon. Her creation, patented in 1914, is better known as the predecessor to the modern bra, and inspired a succession of people to build on her basic design.

Changes in fashion dictated the next stage in the bra's history. A seamstress in New York named Ida Rosenthal (1886-1973) realised that a new shape of

DID YOU KNOW?

- Twenty-three-year-old Berbel Zumner died from bra-related injuries while walking in the town of Vienna. Berbel was very well-endowed, and was wearing a bra with a stretch of metal wiring to hold her in place. Suddenly, a freak bolt of lightning shot through the sky and Berbel was zapped when the lightning found the wires in her bra. She did not survive the experience.
- During the Civil Rights Movement, as free love became increasingly popular in the US, many women publicly denounced their bras as an act of female liberation. The act of 'bra-burning', however, was not actually as popular as the media would have liked us to believe.

bra had to be designed to suit the popular straight-up-and-down flapper dresses of the 1920s. So the undergarment was redesigned and the bandeaux brassiere was produced, which reduced chest size.

Today, there is a vast assortment of colours and styles to choose from. Women in Britain and America own, on average, six bras each, including one strapless and one brightly coloured bra. Some of the different styles of brassiere include:

- Peephole bra
- Training bra
- Push-up bra
- Strapless bra
- Sports bra
- Balcony bra
- Maternity bra
- Minimising bra
- T-shirt bra
- Mastectomy bra

BREAST IMPLANTS

The first recorded breast augmentation surgery was carried out by a Dr Vincenz Czerny (1842-1916) of Heidelberg in Germany in 1895. He removed a lipoma, or a benign fatty lump, from the flank of an actress and transplanted it into her mastectomy site. Over the next few years, surgeons experimented with all manner of augmentation techniques, involving materials such as ivory and glass balls. Most disastrously, paraffin wax was injected into several unfortunate women.

The use of silicon injections was pioneered by Japanese prostitutes immediately after World War II, in an attempt to make themselves more desirable to American GIs. Despite the considerable health risks – the silicon frequently migrated to other parts of the body, sometimes causing death – thousands of women in America followed suit before the practice was banned. In 1962, a safer procedure involving envelopes filled with silicon gel was introduced by plastic surgeons. The first recipient was a 30-year-old mother of six. By the 1990s, saline implants had largely replaced their silicon forebears.

BROWN PAPER BAG

In 1883, a Philadelphia printer and Civil War veteran named Charles Stillwell invented a machine that made paper bags. Sitwell described his creation as the 'Self-Opening Sack, the first bag to stand upright by itself'. The SOS, as it was known for short, was straight-sided, pleated, had a flat bottom and could be opened almost effortlessly. It also collapsed neatly so it was easy to stack. Grocers loved the bags and today billions of them are used annually in the US alone.

BUBBLE WRAP®

The packaging material with air-filled blisters that people can't resist popping was originally designed as an innovative kind of textured wallpaper by engineers Alfred Fielding and Marc Chavannes. They were aiming to create an easy-to-clean plastic wall hanging with that 1950s 'space-age' look. Fielding and Chavannes quickly realised that the product was useless for its intended purpose, but would work remarkably well as a protective wrapping for delicate items. In 1960 they founded the Sealed Air Corporation. The company now has annual revenues in excess of $3.5 billion.

BUBBLEGUM

The world's first bubblegum was made by confectioner Frank Henry Fleer in 1906. Unfortunately, Blibber-Blubber, as Fleer evocatively called his invention, had a number of drawbacks and was never brought to market. It was brittle, unpleasantly sticky, and not elastic enough to recoil back into the mouth when a bubble was blown.

It was one of Frank Fleer's employees, a 24-year-old accountant named Walt Diemer, who finally came up with a workable formula. In 1928, he tried adding latex to Fleer's original Blibber-Bubber recipe, and the product worked a treat. 'It was an accident,' he later confessed. 'I was doing something else and ended up with something with bubbles.' Diemer called his invention Dubble Bubble. Like most of its descendants, the gum was pink, since it was the only food colouring in the factory. He took a lump of Dubble Bubble to a grocery store, and by the end of the first day it had all been sold. Soon Fleer salesmen, trained in the art of bubble-blowing, were demonstrating the product all over America.

A 1940s ad for Fleer's Dubble Bubble gum

BUNGEE JUMPING

Although bungee jumping first achieved mass popularity in New Zealand, it has its roots in a ritual practised by young men on Pentecost Island in the Pacific archipelago of Vanuatu. According to legend, the first bungee jumper was a woman from Bunlap village who ran up a banyan tree after an argument with her husband. When she reached the top, she tied liana vines to her ankles to give her a sense of security. The husband, who was called Tamalie, followed her up the tree, so she jumped. Tamalie did likewise. She survived, he didn't. In time, her example evolved into a ritual called *Gkol*, performed both to guarantee a good yam harvest and to allow the locals to demonstrate their manhood.

Modern bungee jumping was pioneered by members of the Oxford University Dangerous Sport Club, who had seen a documentary about Pacific 'vine jumpers'. On April Fools' Day 1979, they dressed in top hat and tails and launched themselves from the 245-foot-high Clifton Suspension Bridge in Bristol. They used nylon braided rubber shock cords instead of the traditional lianas, and they all survived the stunt.

BURGLAR ALARM

A hoop skirt manufacturer from Boston called Edwin Holmes developed the eletronic burglar alarm in 1852. Using the knowledge of electrics acquired through his day job, Holmes built a device that sounded a bell when the door or window to which it was wired was opened. He also bought up an obscure 1853 patent for 'the development of an electromagnetic alarm' from a Massachusetts man named Augustus R Pope, who then disappeared from the historical record. On 21 February 1858 Holmes installed the world's first electric burglar alarm in his own home in Boston, but he had great difficulty persuading his fellow citizens to follow suit. The main obstacles were a pervading distrust of electricity, and the fact that Boston had a low crime rate. In 1859, Holmes decided to move to New York City on the grounds that this was where most of the nation's burglars 'did their business'. Holmes found a ready market for his alarms among the city's wealthy residents.

U.S. Patent 219,584 December 29, 1970

United States Patent Office

Des. 219,584
Patented Dec. 29, 1970

219,584

BUCKET SEAT

Steven T. McQueen, Los Angeles, Calif.
(14732 Oxnard St., Van Nuys, Calif. 91401)

Filed Dec. 8, 1969, Ser. No. 20,416

Term of patent 14 years

Int. Cl. D6—*01*

U.S. Cl. D15—8

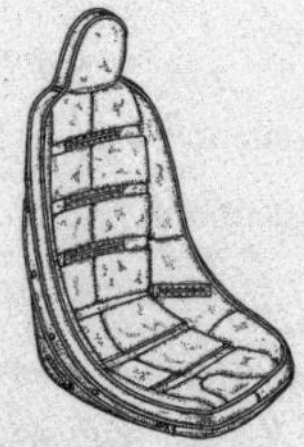

Fig. 1

Fig. 3

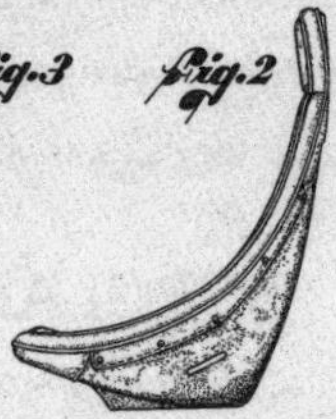

Fig. 2

THE BUCKET SEAT

Amazingly, the cockpit-like bucket seat found in sports and racing cars was invented by the movie star Steve McQueen (1930-1980). The star of *Papillon* and *The Towering Inferno* was a keen driver of racing cars, but found the standard seats unbearably uncomfortable. Therefore, he designed an alternative which followed the contours of the human back. He was awarded a US patent for his invention in 1970. In automobile parlance, the bucket seat is to be contrasted with the bench seat.

C

CAESAREAN SECTION BIRTHS

The operative question regarding caesarean section births is not so much when they were first carried out as when the procedures were first conducted without killing the mothers. In extremis birth attendants have probably been cutting open the wombs of labouring women since prehistoric times, but their chances of survival will have been almost nil. The concept of the caesarean section certainly penetrated ancient myth. The Greek hero Asclepios, who later became the god of medicine, was said to have been removed from his mother's dead body by the god Apollo.

The term caesarean is commonly believed to derive from the name of Julius Caesar, who supposedly came into the world in that manner, but this is unlikely as his mother Aurelia was still alive when he invaded Britain. Caesar did, however, introduce a law decreeing that the operation must be carried out on all Roman women who appeared to be dying during labour. Another possibility is that the word caesarean comes from the Latin verb *caedere*, meaning 'to cut'.

Various prominent medieval figures are known to have been removed directly from the womb at the cost of their mother's lives. They include the Catalan saint Raymund Nonnatus, whose surname meant 'not born', and Robert II of Scotland, the likely inspiration for Macduff, Macbeth's nemesis in Shakespeare's play of the same name. The first recorded example of both mother and child surviving a caesarean section occurred in the Swiss town of Siegershausen in 1500. The local authorities granted a pig gelder called Jacob Nufer permission to perform the procedure on his wife after she had spent several days in unsuccessful labour. Nufer extracted the child at the first attempt and was able to stitch the incision satisfactorily due to his experience with animals. The mother went on to have several more children and the caesarean baby lived to the age of 77.

CAN OPENER

When Nicholas Appert (1752-1841) and Peter Durand developed the world's first tin cans in the early nineteenth century (see canned food)

US Patent 4,360,328 23 November 1982

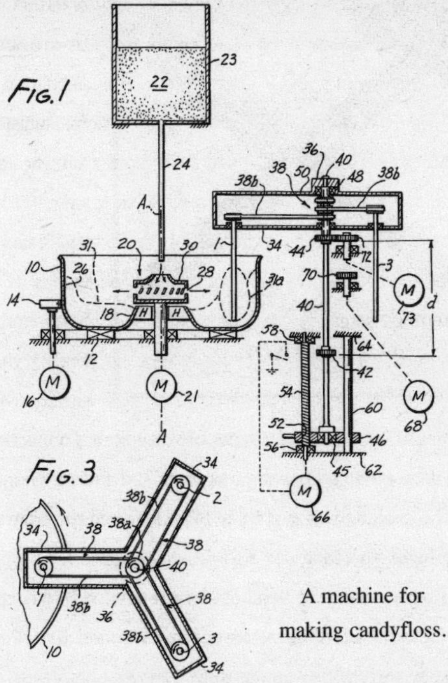

A machine for making candyfloss.

CANDYFLOSS

Candyfloss was invented in 1897 by a pair of confectioners from Nashville, Tennessee called William Morrison (1860-1926) and John C Wharton. 'Fairy Floss', as they initially called their sweet treat, was made by spinning heated sugar in a bowl that had several small holes drilled into its sides. As the bowl rotated, melted sugar was forced through the holes by the centripetal force and hardened into woolly strands when it emerged. Morrison and Wharton knew that they were on to something when they sold 68,655 boxes of candyfloss for 25¢ each at the St Louis World's Fair of 1904.

An early can opener.

they made a startling omission. They forgot to provide a way for people to open them. For almost 50 years, hungry can owners were forced to resort to bayonets, sharp stones or chisels and hammers to get at their victuals. The situation was partially rectified in 1858 when Ezra Warner of Waterbury, Connecticut patented the first dedicated can opener, but users still had to employ plenty of muscle. The curved blade had to be forced around the edge of the rim of a can by hand. The tool was also rather dangerous, so grocers initially insisted that anyone who purchased a tin of food had it opened in the shop by a clerk. Nevertheless, Warner's opener was useful enough to be employed by the military during the Civil War.

The modern form of can opener, which has a cutting edge aligned with a wheel that glides around the top of a tin with ease, was invented in 1870 by William Lyman of Meriden, Connecticut.

CANNED FOOD

Napoleon once famously said that 'an army marches on its stomach'. It is therefore no surprise that he latched on to the fact that the food supplies of his own troops were frequently lost to spoilage. In 1795, Napoleon offered a prize of 12,000 francs to the first person to come up with a reliable way of preserving non-dried foodstuffs for the French Army. The eventual winner was one Nicholas Appert (1752-1841). He collected the reward in 1809 after demonstrating a process in which food was placed in sterilised glass containers, which were then sealed and heated. The following year, a British man named Peter Durand patented a rival technique whereby foods were stuffed into tin-coated steel cans and lids were affixed with solder. Unfortunately, the solder created a new problem: lead poisoning. Victims included the crews of Sir John Franklin's ships the *Terror* and the *Erebus*, who perished in 1845 after becoming stranded in ice off the northern shores of Canada (they were looking for the elusive Northwest Passage to Asia). They had enough canned food to last for three years, but the lead in the solder was the death of them. The lead problem was eventually solved by the invention of the double-folded canning seal.

CAR RADIO

The invention of the car radio was the combined effort of three men: William Lear (1902-1978), Elmer Wavering (1907-1998) and Paul Galvin (1895-1959). The idea began as the amateur brainwave of Lear (the same Lear who invented the Learjet, the world's first mass-produced business

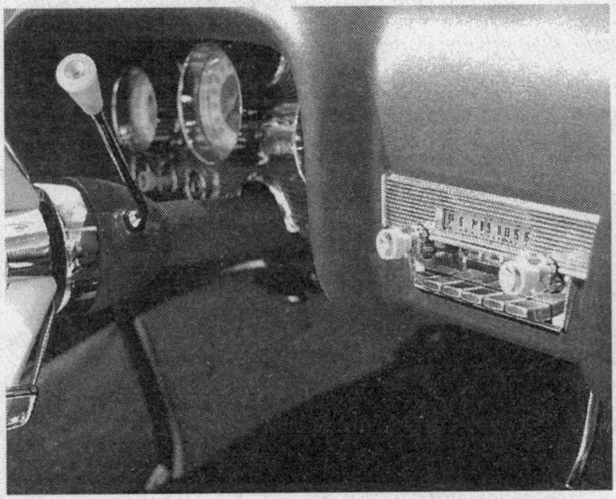

An old-style car radio in all its 1950s chrome splendour.

jet) and Wavering. As a hobby, they played around with ways of installing sound systems in their cars. Not long after this they met radio manufacturer, Paul Galvin, who convinced them to install a radio in his car. At the time, car radios were custom-made and therefore expensive. Galvin saw the potential to mass-produce automobile radios and open up a new market for the Galvin Manufacturing Corporation.

The car radio had a few teething problems in the design stage, but Galvin managed to produce a working model for the Radio Manufacturer's Association Convention in Atlantic City in June 1930. Together he and his brother came up with the label 'Motorola' – inspired by the idea of 'motor' for motion and 'ola' for sound. The brand gradually took off, helped by a successful advertising campaign and soon became a standard fixture in every car.

CAR WASH

The world's first car wash was located in Detroit, Michigan. The Automated Laundry was opened in 1914 by two enterprising residents of the Motor City. It was scarcely sophisticated by modern standards – customers had to leave their vehicles at the premises all day, cars were pushed through the system manually and the brass components had to be moved and polished by hand – but the concept proved popular. The first fully automatic car wash in the UK opened in London in June 1961. It charged a then extortionate 10 shillings.

CARDIGANS AND BALACLAVAS

James Thomas Brudenell (1797-1868), the 7th Earl of Cardigan, was a mediocre soldier – he was responsible for the Crimean War's disastrous Charge of the Light Brigade, immortalised in Tennyson's poem – but he had a flair for knitted garments. He is credited not only with the invention of the woolly waistcoat that bears his name, but also with the balaclava. The sinister hood with holes for the eyes and mouth was first worn by British troops at the Battle of Balaklava in 1854 (the engagement during which

The charge of the light brigade. Not Cardigan's finest hour, though the Battle of Balaklava did give rise to the headwear of the same name – a somewhat better legacy!

the fateful charge took place). Cardigan designed the garment to protect his soldiers from the bitter Crimean winter.

CATSEYES®

One night in 1933, a 43-year-old road contractor named Percy Shaw (1890-1976) was driving back to his home in Halifax, Yorkshire when he noticed his headlights illuminating some objects at the side of the road. They turned out to be reflectors on a roadside poster and inspired Shaw to develop a reflecting road stud. That is one of several stories of how Shaw came up with the idea; another is that the illuminating objects were indeed the eyes of a cat. Nevertheless, his design incorporated rubber housings into which the reflective lenses were forced when a car passed over them. (You can have a lot of fun squirting unsuspecting friends by jumping on a catseye after a rain shower.)

On 4 April 1934, the year in which he patented his invention, Shaw laid the world's first row of catseyes at a notorious accident black spot near the city of Bradford. During World War II, the British government placed a huge order to help the forces cope with blackouts. In 1947, the Transport Minister and future Prime Minister Jim Callaghan ordered catseyes to be installed in roads throughout the country. When Shaw died in 1976 he was a very wealthy man.

CELLOPHANE

The clear transparent plastic known as cellophane might never have come into existence were it not for the clumsy eating habits of an anonymous man from Switzerland. One day in 1900, an aristocratic textile engineer called Jacques E Brandenberger (1872-1954) was dining in a restaurant when the gentleman on the adjacent table upset a glass of red wine. As he watched the harassed waiter replace the tablecloth, Brandenberger was struck by an idea. If a transparent, flexible and waterproof film could be applied to material, he reasoned, such incidents could be avoided. So he went home and started to experiment with chemicals and scraps of cloth.

The moment of truth came when he immersed some cloth in rayon (liquid viscose, made from cellulose extracted from plants). The result was rigid, and therefore utterly useless from a catering perspective, but Brandenberger noticed that the coating was easy to peel off. When he did this, he found he was holding a transparent sheet of film. It was made of cellulose and was *diaphane* ('see-through' in French), so he decided to elide the words and call the new material 'cellophane'.

It took Brandenberger eight years to develop the machinery to produce cellophane reliably and efficiently, but as soon as he had done so, he knew he had a commercial winner on his hands. He quickly secured patents for both the material and the manufacturing process, and before long, chic shops began to use cellophane as a wrapping for upmarket goods. For the first time, customers could see what they were buying through the packaging. The days of brown paper and string were numbered.

CELLULAR PHONE

'People want to talk to other people – not a house, or an office, or a car. Given a choice, people will demand the freedom to communicate wherever they are, unfettered by the infamous copper wire. It is that freedom we sought to vividly demonstrate in 1973.'

So said Dr Martin Cooper – former general manager of Motorola Communication Systems and latterly co-founder of ArrayComm, Inc – in 2003, the 30th anniversary of his success as the inventor of portable mobile phone technology. Born in Chicago, Martin Cooper earned a degree in electrical engineering and then served in the US Navy. After a brief stint at a telecoms firm, he was hired by Motorola in 1954 and over the next two decades dedicated himself to developing portable technology. As he was completing the last stages of his tenacious research to produce mobile telecoms, rival firm Bell produced a radio-wave telephone system that they launched in 1969 as a payphone network on the New York to Washington train line. The Bell system adopted the name of the train service – Metroliner – and even filed a patent for it. This was approved in 1972. Fearing a case of innovation leapfrog, Cooper and co knuckled down.

An early brick of a phone. No you can't take photos with it.

Strangely enough, anyone who was researching mobile telecoms at the time had a common enemy: the US Federal Communications Commission (FCC). The government body had long been reluctant to hand out extra frequencies to the various developers of mobile phone technologies, but the persistence of those involved and the obvious commercial benefits finally conspired to release a raft of wavebands from the FCC's iron fist.

Then, in 1973, Cooper broke through with a device that would change the world's telephonic landscape: the prototype Motorola DynaTAC. Roughly the size of a brick, the bulky handset – reminiscent of the field telephones in countless Vietnam films – was nine inches long, contained 30 circuit boards and weighed two and a half pounds; not so much an executive toy as a dumbbell. It also took 10 hours to charge, in return for which it provided a meager 35 minutes' talk time. However, such limitations did little to hamper the DynaTAC's maiden mission, which may rank as the smuggest phone call ever made. In a 1999 interview, Cooper revealed that his first use of the DynaTAC was to call up Bell's head of research, Joel Engel, to tell him how well the Motorola project was coming along.

Cooper also amused himself by eliciting the first few public reactions to the new shape of telephony: 'As I walked down the street while talking on the phone, sophisticated New Yorkers gaped at the sight of someone actually moving around while making a phone call. Remember that in 1973, there weren't cordless telephones, let alone cellular phones. I made numerous calls, including one where I crossed the street while talking to a New York radio reporter – probably one of the more dangerous things I have ever done in my life.'

CHEESE

Cheese-making was already old when Homer referred to it in *The Odyssey.* Cheese is mentioned several times in the Old Testament. In 1 Samuel 17:18, for instance, King David presents 10 curd cheeses to the captain of his army ahead of his battle against Saul. The archaeological evidence stretches back even further. Sumerian reliefs from 5,500 years ago have been found depicting the curdling of milk in special vessels. Cheese was probably invented, or rather discovered, by nomads from Central Asia or the Middle East. All that was necessary was for milk to be stored in an animal stomach – pouches made from innards have long been used in desert regions. At some point in prehistory, an unknown individual must have opened such a container and found that the milk had been transformed into a tasty semi-solid. Thousands of years later scientists identified the agent responsible for the curdling as rennet, a naturally occurring enzyme found in the stomachs of ruminants.

CHESS

Several nations claim to have invented chess, including India, China, Iran (Persia), Arabia and Uzbekistan. In many ways, the debate is pointless, because the game did not suddenly arrive in its modern form, but slowly evolved. The pawn, for example, only acquired its current power to make an initial move of two squares in Renaissance Italy. What is undoubtedly true, however, is that the names for chess in the languages of many of the main candidates, including Persian and Arabic, are derived from the Sanskrit word *Chaturanga*. *Chaturanga*, which means 'four parts', was played by four players on a board with 64 squares. The pieces consisted of a king and his general in the centre, plus representatives of the four main divisions of the ancient Indian army: infantry, war elephants, chariots and cavalry. This strongly implies an Indian origin. This notion is supported by one of the earliest unambiguous references to the game, written around 600AD in Persia. It states that a *Shatranj* set (the Persian word for chess) was presented to the Emperor Naushirawan (531-579) by an Indian ambassador.

Interestingly, the term 'checkmate' is derived from the Persian *shah mat* ('the king is ambushed').

CHEWING GUM

Indigenous peoples have chewed various substances since the dawn of time. The Inuit of the Arctic, for example, are fond of chewing narwhal skin. They call it *muktuk*, and it is a valuable source of vitamin C in an environment devoid of fresh vegetables for much of the year. The Ancient Greeks liked to chomp on the resin of the mastic tree, and Native

One of Thomas Adams' chewing gum flavours.

Americans have long chewed spruce tree resin, a habit they passed on to early New England settlers. This was the basis of the first commercial variety, which John B Curtis (1827-1897) began selling in 1848 as 'State of Maine Pure Spruce Gum'. The ancestor of modern chewing gum, however, is latex from the Sapodilla tree, which the Mayans of Southern Mexico and Central America have been harvesting since the second century AD. Known as *chicle*, this substance came to the attention of a New York entrepreneur named Thomas Adams (1846-1926) during the 1860s, via his acquaintance with the exiled Mexican General Antonio Lopez de Santa Anna (1794-1876). At first, Adams tried to convert chicle into bicycle tyres and other synthetic rubber products, but he drew a blank. One day in 1869, he put a spare piece of stock into his mouth in exasperation. It tasted good. Inspired, Adams built the world's first chewing gum machine, and in February 1871, he began selling the product through drugstores. One of his early gums, the liquorice-flavoured Black Jack, is still popular today.

CHOCOLATE

The cacao plant which yields the beans used to make chocolate is believed to have been first domesticated by the Olmec tribe of southern Mexico between 1,500 and 500BC. The chocolate habit then spread to the neighbouring Mayans. Cacao residue dating from 460AD has been found in a ceramic pot from Rio Azul, a Mayan site in Guatemala. The next great

'High as the Alps in Quality'
A 1906 ad for Daniel Peter's Swiss milk chocolate.

chocaholics were the Aztecs. They used it to make a red-dyed drink called *xocoatl* ('bitter liquid') which was believed to bestow wisdom and power on those who consumed it. Hernando Cortés, who conquered Mexico in 1519, recorded that the Emperor Montezuma (1480-1520) drank up to 50 goblets of xocoatl per day. When the conquistadors brought cacao beans back with them to Spain, the locals took to sweetening the drink with sugar. They also decided that it was more palatable when drunk warm. Chocolate remained primarily a beverage until 1847, the year in which the British firm Joseph Fry & Sons manufactured the first edible bars. Milk chocolate was invented by Daniel Peter (1836-1919) of Vevey in Switzerland in 1876.

CHRISTMAS CRACKER

The origins of the Christmas cracker date back to the Victorian era and a man called Tom Smith (c1820-1881), who was a confectionery assistant in London. Inspiration first struck him after a trip to Paris in 1840 where he was very taken with the French tradition of wrapping sugared almond bonbons in tissue paper.

ORIGINAL FACTS

Did you know that the world's largest cracker measured 63.1 metres long, and 4 metres in diameter? It was built by Ley Hill School in Buckinghamshire, England, and was filled with 300 balloons to raise money for charity.

He began to replicate this idea back home, introducing love mottos into the sweets he sold to boost sales. The idea for the bang occurred to him when he was staring at the fire one night and listening to the crackle. A 'crackle' would add the excitement he was looking for. He just needed to work out the safest compound to create the effect. Eventually he managed to generate a 'pop' from the friction when the wrapping was broken. Interest in his noisy bonbons tooks off and Smith took to refining his idea, incorporating a surprise gift and renaming them 'Cosaques'.

Smith soon had a thriving business going. The company expanded so much that the premises moved to Finsbury Square, where it remained until 1953. When he died he left the business to his three sons, Walter, Tom and Henry, who set to reinventing the concept. Contemporary crackers continued to use Smith's love mottos, but these were gradually phased out to include puzzles, cartoons and today's corny jokes. Walter also introduced the paper hat. Today, the Tom Smith factory still continues to churn out crackers for an international market.

CIGAR

The word 'cigar' is thought to derive from *sikar*, the Mayan word for smoking. The notion that the Mayans were the first to smoke tobacco in the form of cigars is supported by an image on a ceramic vessel dating from 900AD found at Uaxactún in Guatemala. The first smokers of European descent appear to have exclusively used pipes. Cigars made with Cuban tobacco became popular in Spain during the eighteenth century. It was from Cuba that the habit spread to the US. In 1762, Israel Putnam brought a selection of cigars back to his home in Connecticut after serving time on the island as an officer in the British Army. The first American cigar factories sprang up in the Hartford area soon afterwards.

CIGARETTE

In November 1492, a member of Columbus's entourage named Rodrigo de Jerez became the first European to smoke tobacco. He described the natives as 'drinking' smoke from items made 'in the manner of a musket formed of paper'. As they were wrapped in palm or maize rather than tobacco leaves, they sound more like cigarettes than cigars. De Jerez evidently partook: he returned to Europe a committed smoker and was imprisoned by the Spanish Inquisition for frightening his neighbours.

For the first few centuries of its existence in the West, people smoked tobacco in the form of pipes or cigars. The earliest reference to it being rolled in paper concerns an Egyptian gunner at the siege of Acre in 1832. Hand-rolled cigarettes were first manufactured in France in 1843, and in Britain in 1856. They remained expensive luxuries until the early 1880s. In 1881, an 18-year-old Virginian named James Bonsack (b1862) patented an automated machine that produced 200 cigarettes per minute. Skilled hand-rollers were pushed to make four in the same period. The price of cigarettes plummeted and annual sales leapt from a peak of 500 million prior to Bonsack's invention to 10 billion by 1910.

Before the days of health warnings, cigarettes were promoted as stylish and aspirational.

COAT HANGER

One of the puzzles of modern life is how we (and dry cleaners) ever survived without the coat hanger. Until the twentieth century, people had to suffice with the good old-fashioned coat hook. But there are just never enough coat hooks around when you need them, and this was the problem that Albert J Parkhouse (1879-1927) set out to solve. Parkhouse was an inventor at the Timberlake Wire and Novelty Company in Jackson, Michigan, which made wire novelties, lampshades and other wire items for clients. In a moment of inspiration, Albert took a piece of wire, bent it into two large hoops either end, attached another piece in the centre to act as a hook, and voila! The wire coat hanger was born.

He continued to refine the idea and soon his colleagues were all asking for replicas. John B Timberlake, his employer, realised the idea had serious potential and in 1904 he applied for a patent on the wire coat hanger. Parkhouse was named as the inventor, but it is not known whether he profited from his invention. Between 1900 and 1906, over 189 different patents were granted on different versions of the 'garment hangers' worldwide. Wire was used as the main structure, sometimes combined with wood or fabric, depending on the garment. Over the years the hanger saw many different incarnations of Parkhouse's original idea; speciality hangers which allowed belts to be hung from them, travelling collapsible hangers and spring-coiled ones. Parkhouse had opened up a whole world of hanging possibilities.

COCKTAIL

The earliest known written reference to the term 'cocktail' as a drink based on spirits with other spirits and/or other additives goes back to an early American magazine called *Balance, and Columbian Repository*, published in May 1806.

'Cocktail is a stimulating liquor, composed of spirits of any kind, sugar, water, and bitters – it is vulgarly called bittered sling and is supposed to be an excellent electioneering potion.'

The origin of the cocktail is highly contested and may never be fully fathomed. Most accounts agree that the drink is rooted in American culture, but one popular story traces the origins back to a Creole man from a wealthy French family named Antoine Amedee Peychaud (b1803). Peychaud and his wealthy plantation owners fled their home in the French controlled portion of the island of Hispaniola during the slave uprisings of 1793. He settled in New Orleans and set up an apothecary in the French quarter. One of the family recipes he took with him was an old concoction for the compounding of a liquid tonic known as 'bitters'. The bitters were meant to be good for any ailment and Peychaud served it with cognac brandy to his friends who visited the apothecary. The fame of his tonic began to spread and soon New Orleans coffee houses (as drinking houses were then called) were serving French brandy with a dash of Peychaud's magical bitters. Peychaud had a unique way of serving his bitters, in a double egg cup called a *coquetier* (French for eggcup). The story goes that the word became corrupted into the present day 'cocktail'.

A more elaborate story tells the tale of Betsy Flanigan an innkeeper of Yorktown in America. Betsy had an ongoing dispute with her neighbour, who kept chickens. One day, having served a meal to her regulars, she produced a round of drinks decorated with chicken feathers, and announced that the meal they had just consumed was made from the neighbour's stolen chickens. The drinking went on into the night, to cries of demand for more 'cocktails'!

Other possible explanations include:

- A contemporary reference to **'cock-tailed' horse** – meaning a horse of mixed blood – was applied to the drink because it was also a mixture.
- **'Cock-ale'**, a meal of bread fortified with mixed spirits, was served to fighting cocks – and spectators – before a contest.
- ***Cock tailings*** were the dregs of casks of spirits, which would be drained out through their cocks (spigots), mixed together, and sold as a cheap drink.

These are just a few of the more popular stories and evidence for each of them is equally flimsy. Although the tale of Peychaud and the *coquetier*, sounds the most convincing, this is confounded by the dates, as it turns out that Peychaud opened the apothecary shop in 1838, long after *Balance, and Columbian Repository* defined the cocktail, which casts doubt on the suggestion that Peychaud thought of it first. It seems that the cocktail is destined to remain a conundrum – perhaps providing another name for a cocktail recipe?

COFFEE

The coffee plant *Caffea arabica* is native to the highlands of Southern Ethiopia. According to a legend told in the Kaffa province from which the beverage takes its name, its properties were discovered by a goat-herder named Kaldi towards the end of the first millennium. One day he noticed his animals becoming hyperactive after eating the red berries from the wild plant. He ate some himself and experienced a similar effect. A passing monk scolded Kaldi for eating 'the Devil's fruit' but was eventually persuaded to try some himself. He realised that the berries could help him and his brothers stay awake during prayer. Members of the Galla tribe of Southern Ethiopia are also known to have eaten coffee beans wrapped in fat to sustain them during hunting trips.

The first people to use coffee beans to make a beverage were Yemeni Arabs in about 1000 AD. The habit eventually spread to Turkey, where the practice of roasting the beans ahead of use was developed, and thence to Italy via Venetian merchants.

COMPUTER DISK (CD)

Where would we be today without 'Optical digital data storage' technology? This was the name of the technology behind the CD, as defined by its inventor Jim Russell (b1931). Russell, however, is not a billionaire; in fact he still works in his basement to supplement his pension from his employer, global science and technology firm Battelle.

US Patent 3,501,586 17 March 1970

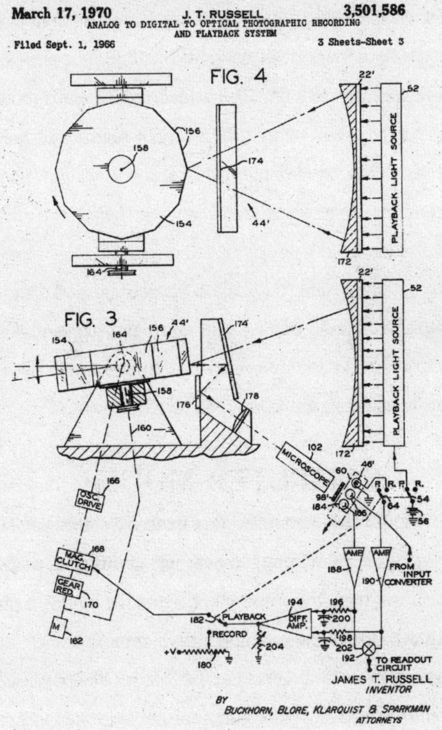

Part of Jim Russell's original compact disk patent application.

Jim Russell was obsessed by music and was looking for a way to get a better quality of sound than vinyl or audiotape could provide. His preferred method of listening was to use a cactus needle on his records, hand-sharpened after every use, but even this wasn't good enough and the records always deteriorated. He wondered if using a laser beam would create a sharper sound. In 1965, using a computer that changed analogue data into digital, he devised a light-sensitive system of reading the data and synchronising it to create more than ones and zeros. Unfortunately,

Battelle's verdict was, 'if it was any good, IBM would have already invented it', and they ceased funding his work. Battelle was unaware of the vast potential of the system of storing data that Russell was working on, and sold the patents for a mere US$1 million.

However, Russell pursued the idea and eventually a man named Eli Jacobs began funding research in 1973. In 1974 Russell had built his first prototype. They demonstrated it to Philips and Sony but they weren't impressed. However, three years later, Sony and Philips joined together to develop digital-audio-recording and released their CD technology in 1982. Russell knew that they were stepping on his patents, and after a lengthy battle, ORC (the company that bought the rights to Russell's patents) settled for $30 million in 1988. Nevertheless, to this day, Philips maintains that Piet Kramer invented the CD.

COMPUTER MOUSE

Invented in 1963-4 by American researcher, Dr Douglas Engelbart, the mouse was a more accurate means of 'pointing and clicking' on a computer screen than the more cumbersome method of shunting a cursor around with the arrow keys and hitting 'return'.

Dr Engelbart wrote in his original outline that his device would be: 'held like a pencil and, instead of a point, has a special sensing mechanism... The signals which this reading stylus sends through the flexible connecting wire to the writing machine are used to determine which characters are being sensed... An information-storage mechanism in the writing machine permits you to sweep the reading stylus over the characters much faster than the writer can type.' A late-2001 interview with Engelbart shows that his earliest designs would hardly have qualified for the name of 'mouse'. 'We had a big heavy tracking ball,' he said. 'It was like a cannonball.' Other stages of development also proved to be false dawns. 'We had several gadgets that ended up with pivots you could move around. We had a light panel you had to hold up right next to the screen so the computer could see it. And a joystick that you had to wiggle around to try to steer things.'

US Patent 3,541,541 17 November 1970

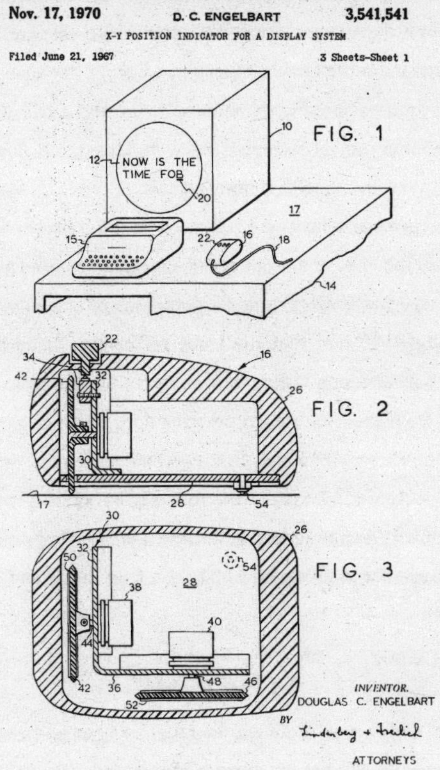

Part of Engelbart's patent application for the first computer mouse.

Despite its success, he was never entirely thrilled with the name that his creation acquired: 'One of the members of the team nicknamed the device a mouse and it caught on. We thought that when it had escaped out to the world it would have a more dignified name, but it didn't.'

Engelbart now runs the advanced technological think-tank, The Bootstrap Institute, where his ongoing project is to develop methods of networking human brainpower. 'If we don't learn how,' he says, 'it's very likely that our society will crash.'

CONDENSED MILK

Gail Borden Jr (1801-1871), born in Norwich, New York, was a serial inventor. His early schemes included a 'locomotive bathhouse' for women wishing to swim in the Gulf of Mexico in privacy, a 'terraqueous' vehicle capable of functioning on both land and water, and a dehydrated meat biscuit. The latter nearly bankrupted him, but in the 1850s, Borden hit the jackpot with the invention of condensed milk.

While Borden was sailing back to the US from Britain, several children on board his ship died as a result of drinking contaminated milk. Borden had witnessed the Shakers using vacuum pans to condense fruit juice and it occurred to him that the same technology could be used to produce a safe and long-lasting milk product. The keys were to remove the water from the milk and to be scrupulous about hygiene. Borden patented the idea in 1856 and, after two failed attempts, established a successful factory in Wassaic, New York. His invention proved timely: during the Civil War, the Federal government ordered huge quantities of Borden's Eagle Brand Condensed Milk as a field ration and its success was secured.

CONDOMS

Evidence of early condom use was found in an Egyptian painting dating from around 1,350BC, but the purpose of the penile sheath worn by the man portrayed is uncertain. Its function was probably ritualistic rather than contraceptive. Between 100 and 200AD, scenes showing men wearing condom-like devices were painted on the walls of a cave in Combaralles in the south of France. The first written description of the condom appeared in *De Morbo Gallico* ('On the French Disease', ie syphilis), a posthumously published work by the Italian anatomist Gabriello Fallopius (1523-62). In the book he claimed that prophylactics of his own design issued to 1,100 men had combated the disease in 100% of cases. Fallopius's condom consisted of an eight inch sheath made of medicated linen which was tied into place with pink ribbon.

The word 'condom' is thought to have come into use during the mid-seventeenth century. The name is often ascribed to a Dr Condom or Quondam, who was either a physician during the English Civil War or court doctor to King Charles II depending on the account. Both stories are probably apocryphal. The term is more likely to be derived from the Latin *condom*, meaning 'receptacle'. The oldest surviving examples were discovered in Dudley Castle in the British Midlands among the rubble in a lavatory that was filled in around 1647. They are made from animal gut finely stitched together at the 'business end'.

In Europe, animal gut remained the preferred condom material for several centuries. The Chinese used oiled silk paper and Japanese versions were made from leather and tortoise skin. It was only after Charles Goodyear (1800-1860) invented the vulcanisation process in 1839 that the first 'rubbers' emerged. Victorian condoms had to be washed before and after deployment and were used until they wore out. This less-than-satisfactory state of affairs persisted until latex condoms were introduced during the 1930s.

CONTACT LENSES

The concept of the contact lens was first propounded in 1508 when Leonardo da Vinci sketched and described a rudimentary example in one of his notebooks. The device consisted of a short tube with a flat lens at one end. It was to be filled with water and held against the eye via the open end. The water would then correct faulty vision by bending the light from the object under inspection in much the same way as the curved lenses of today.

The first true contact lenses were made by a doctor named Adolf Eugen Fick (1829-1901) in 1887. They were made of heavy brown glass and covered the entire eyeball. Fick initially tested his invention on rabbits before donning a pair of the lenses himself. They were far from comfortable, but they worked. Glass lenses remained the norm until 1936, when a New York optometrist named William Feinbloom introduced a hard plastic version. 'Soft' contact lenses were developed during the 1960s.

CORKSCREW

The exact origin of the corkscrew is unknown, but appears to be an English creation and dates back to the 1600s. In the second half of the seventeenth century, the English started to mature their wine in bottles rather than casks. Glass manufacturing was steadily improving and the bulbous flask-type container was replaced with a more cylindrical shape, which allowed them to be stored on their sides. This meant that the wine stayed in contact with the cork and did not dry out. A device was therefore needed to decork.

It is thought the inspiration came from a tool called a bulletscrew or a gun worm used to extract jammed bullets from rifles. The original design was like the waiter's corkscrew; a basic T-shape. The only problem with this being the force with which it had to be pulled out, as many a struggling waiter will tell you. Before the age of factories and heavy machinery, it was only blacksmiths and jewellers who were skilled enough with metal-working to make corkscrews. The Industrial Revolution and the development of machinery meant that better methods of de-corking could be developed, leading to the more modern plunge-type device.

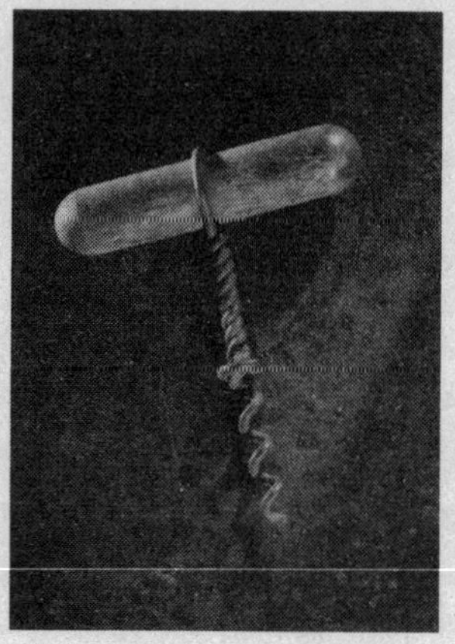

The original T-shaped corkscrew design.

CORNFLAKES

Cornflakes owe their existence to a vegetarian diet designed by a group of Seventh Day Adventists to dampen the 'animal passions'. They included in their number Dr John Harvey Kellogg (1852-1943), head of the Battle Creek Sanitarium in Battle Creek, Michigan. He was a great believer in sexual abstinence and insisted that his patients ate a bland diet to help them resist temptation. One day in April 1894, Dr Kellogg and his brother Will Keith Kellogg (1860-1951) left some cooked wheat unattended and returned to find it had gone stale. Being frugally minded, they attempted to salvage it by rolling it into dough, but instead of a continuous sheet, the roller spat out flakes. The brothers decided to toast them and feed them to their patients. They went down a treat, particularly when eaten with milk. In 1894, the pair registered a patent for the product, which they initially called 'Granose'. Two years later, after experimenting with other grains, Will created the cornflake.

In 1906, Will decided to leave the sanitarium to concentrate on breakfast cereal full time. He set up the Kellogg Company to manufacture and market his cornflakes. Heretically, from the point of view of his brother, he added sugar. The cereal was an immediate hit, and cornflakes were his first marketed product.

COTTON

Cotton has been cultivated since prehistoric times by cultures all around the world. One of the oldest discoveries was in a cave in Mexico and similarly ancient finds have been found in Mesoamerican civilisations. It was also used in ancient Indian, Egyptian and Chinese cultures. So where does cotton come from? Sadly it doesn't hang from a tree in convenient threads, but comes from shrubby plants and clings in balls of downy white fibre to the seeds, known as cotton bolls. It is these fibres that are then spun into threads.

Although the exact origin of its cultivation is unclear, its development in Europe is more traceable. The Moors introduced the cultivation of cotton into Spain in the ninth century, and it is likely to have spread from Asia to the Mediterranean through Arab traders. It began appearing in Britain in the seventeenth century as the East India Company brought back rare fabrics. However, importation was initially banned to protect the British wool industry.

Cotton production began in the American colonies in the 1600s. They had the potential to produce a huge amount of cotton, but lacked the technical knowledge. It took the expertise of an English mill worker, Samuel Slater (1768-1835), to introduce the first working mill, when he emigrated to America in 1790 and built a mill from memory. It was still a labour-intensive process though, as the lint (cotton fibres) was separated from the seed by hand.

A man named Eli Whitney (1765-1825), who was keen to get into the business, therefore looked into ways of improving this process, and in 1793 invented the cotton 'gin', his nickname for engine. This speeded up the process and dramatically increased productivity. Cotton became the chief cash crop and the slave labour economy of the southern states, and a key economic cause of the Civil War. But with the end of slavery and the spoiling of the soil due to the war, the cotton belt was pushed into the western US states.

Today, cotton is still a major industry in America and until recently America was the world's leading supplier of cotton, second only to Britain. However, China and India are now the leading cotton manufacturers. Even with all the man-made fibres we have today, it is still one of life's essentials, and looks set to continue for many more generations.

CRISPS

The thin salted crisp, known as a 'potato chip' in the US, was invented by an exasperated Native American chef named George Crum (1822-1914). One day in 1853, a diner at Moon Lake Lodge in Saratoga Springs, New York sent his order of French fries back to the kitchen on the grounds that they were too thick and soggy and needed more salt. Crum shrugged his shoulders and prepared a thinner, crispier batch. The man was still not satisfied. This time, Crum was seriously annoyed. He decided to punish the customer for his pedantry by preparing the thinnest, crispiest and saltiest potato slices he could possibly make. To his amazement, the fussy diner was delighted with them. When other customers started to request the new delicacy, the Moon Lake Lodge added 'Saratoga Chips' to its menu, and the crisp was born.

CROSSWORD PUZZLE

The world's first published crossword puzzle appeared in the *New York World* on 21 December 1913. It was diamond shaped, there were no black squares and the first answer was provided to get the reader started. The first clue, which would now be described as '2 across', was 'What bargain hunters enjoy'. (The answer was 'sales'.) The format was devised by an immigrant from Liverpool named Arthur Wynne (1862-1945), who drew his inspiration from a Roman game called 'magic squares'. Wynne initially called his invention a 'word-cross'.

CUCKOO CLOCK

Anyone who has seen the classic movie *The Third Man* (1949) will remember Orson Welles's character Harry Lime's scathing verdict on the consequences of peace:

'In Italy for 30 years under the Borgias they had warfare, terror, murder, and bloodshed, but they produced Michelangelo, Leonardo da Vinci, and the Renaissance. In Switzerland they had brotherly love – they had 500 years of democracy and peace, and what did that produce? The cuckoo clock.'

But Lime was wrong. The cuckoo clock was actually invented in the village of Schönwald, near Triberg, in the Black Forest of Germany. In 1738, a local clock-maker named Franz Anton Ketterer (1676-1749) adapted a Dutch timepiece to imitate the call of the cuckoo via a system of bellows and whistles. Ketterer's invention sparked something of a craze: by 1808 there were 688 cuckoo clock makers in the districts of Triberg and Neustadt alone. The majority of cuckoo clocks are still manufactured in Germany, but the notion that they are fundamentally Swiss persists. *The Third Man* is partly responsible but Mark Twain must also take a share of the blame. In his novel *A Tramp Abroad* (1880), he wrongly described the Swiss town of Lucerne as 'the nest of the cuckoo clock'.

D

DARTS

The game of darts has a rich history that dates back to medieval England, although the exact details of how the game came about are unclear. Historians believe that those teaching archery shortened the arrows and had students throw them at the upturned bottom of wine barrels. When there were not enough wine barrels around to use, a darts player brought in the cross-section of a moderate-sized tree. The rings on the tree provided the segmentation, which became more pronounced as the board cracked, evolving into the modern dartboard. The game soon became popular among the upper classes and King Henry VIII (1491-1547) is reported to have been a keen player.

Darts became increasingly popular in the nineteenth century when the rules began to form into what they are today, although not without a certain amount of trial and error. An early version of darts, called Puff and Dart, used a blowpipe to fire the dart at the target. In 1844, during a game at a London pub, a player made the mistake of sucking rather than blowing and swallowed the dart. He died a few days later.

The modern dartboard evolved from the cross-section of a tree trunk, used as a target when no wine barrels were available.

It was not until the twentieth century that darts became a serious pub game known as 'Dart and Target' according to *Lawful Games on Licensed Premises* (1904). This consisted of a board of numbered coloured circles, without doubles and trebles. The highest score was the bullseye and the lowest score was at the edge. Around 1925, brewers began to organise darts leagues. In 1937, the game saw a massive boost in popularity after the king and queen turned their hand to throwing a few darts on a tour of a social club in Berkshire. *The Sunday Chronicle* reported that 'the queen has made the women of Britain darts-conscious'. Today, darts is played internationally and is currently being considered for Olympic status.

DENIM JEANS

In the 1970s, trade papers commented: 'Jeans are more than a make. They are an established attitude about clothes and lifestyle.'

Today, they come in countless shapes and sizes: long and lean, flared, skinny, hipster, boot-cut, cut-off – a far cry from their functional, humble origins as work overalls. Jeans are most famously associated with America, but their heritage appears to date back to seventeenth century Europe. Denim is actually one of the world's oldest fabrics. It is thought to originate from the French term *Serge de Nîmes* – from the town of Nîmes in France. However, this fabric was a mixture of silk and wool, whereas denim has always been made from cotton. A similar fabric bearing the same name was also in existence in England. It is possible it was mistranslated in England into 'de-Nim', but it does not look as though we will ever get to the bottom of its origin. Jean was a completely different fabric altogether. Whereas denim was made of one coloured thread and one white thread, jean was two threads of the same colour woven together.

Denim production in the US begun in the eighteenth century, as manufacturers sought to become independent from foreign suppliers. A factory in Massachusetts began weaving denim and jean. Denim and

US Patent 139,121 20 May 1873

Part of Jacob Davis' original patent application.

jean differed in how they were used, though. Denim was more of a coarse cotton used for work overalls, whereas jean was less sturdy and reserved for more tailored trousers. Those engaged in manual labour tended to wear overalls made of blue denim, whereas those in more blue-collar jobs wore more finely tailored jeans.

So how did modern-day jeans come to be called jeans when they were actually made of denim? The answer partly lies in the name synonymous with jeans – Levi Strauss (1829-1902). Born Leob Strauss in Bavaria, he moved to New York with his mother and two sisters in 1847. Here he helped his brothers run a wholesale dry goods business. By 1850, he had changed his name to Levi and applied for US citizenship. He decided to move to San Francisco and set up his own branch of the business. Among other items, he sold work pants.

These caught the eye of a Nevada tailor called Jacob Davis who approached Strauss with the idea to strengthen his pants using metal rivets. Davis had already started to do this for his customers and it had proved incredibly popular. However, he did not have the finances to patent the idea, so proposed that he and Strauss go into partnership. Strauss was quick to catch on to this moneymaking potential, and in 1873 agreed to patent the idea of 'improvement in fastening pocket-openings'. He called them 'work overalls', the old term for jeans and chose to make them from denim, as this was more appropriate for work wear.

The more durable riveted style became a popular seller, and by the 1920s, it was the most successful selling work pant in the Western states. With the evolution of the cowboy movie in the 1930s, Levi's were given cult status, immortalised on screen by the likes of the swaggering John Wayne and Gary Cooper. Gradually Levi's became less associated with work wear and more associated with casual wear. Post World War II, it was clear their client base was changing dramatically as teenagers and college boys began to buy them. These younger consumers coined the name 'jeans', and by the 1960s, Levi's re-branded its work pants 'jeans', as the popularity of the term proved so strong. Today, jeans are a universal phenomenon which looks set to continue.

DENTAL FLOSS

As archaeological finds have confirmed, people have been using fibrous material to get at maddening items stuck between their teeth since prehistoric times. The first man in the modern era to promote flossing as part of a daily hygiene routine was a New Orleans dentist named Levi Spears Parmly (1790-1859). In 1819, he published a book entitled

A Practical Guide to the Management of Teeth in which he claimed that complete dental health could be achieved with three tools: the tooth brush, tooth powder and lengths of waxed silken thread. Parmly wrote that: 'a thread passed between the teeth after every meal will save more teeth from decay than all the brushes and powders that can be used where the waxed thread is neglected.' Silk remained the standard flossing material until it was replaced by nylon during the 1940s.

DENTURES

The first people known to have developed false teeth are the Etruscans, who preceded the Romans as the dominant tribe in central Italy. A surviving Etruscan jaw from around 700BC has gold bands soldered around the owner's remaining natural teeth, together with several implanted human teeth and one ox tooth. Other specimens from the region show that wealthy Etruscan women went in for gold bridgework, while others replaced their natural incisors with substitutes made from the same metal.

The most prominent early American denture wearer was George Washington (1732-1799). In 1789, a dentist called John Greenwood (1760-1819) made him a lower set from a hippopotamus jaw inlaid with eight human teeth. By 1798, all the first president's teeth had fallen out, so Greenwood made him a full set. The upper denture consisted of ivory teeth mounted in a gold palate. It was connected by springs to a bottom denture carved from a single block of elephant ivory.

DIGITAL WATCH

The first electric watch was produced by The Hamilton Watch Company of Lancaster, Pennsylvania in 1957, and was the precursor to the digital watch. Its benefits were hard to see at first; as well as being bulky, the electric watch was no more accurate than a conventional timepiece, and although it did not need to be wound, the electric contacts became worn quite quickly. The Accutron watch, made in 1960 by American manufacturer, Bulova, used transistors instead, which increased the lifespan, as well as special resonating tuning forks as a way of regulating the timekeeping. This product was used on the NASA Mercury and Apollo spacecraft. Max Hetzel, the man behind both these innovations, moved to Omega where he honed the technology; Omega watches were also widely used by NASA astronauts.

The success of these designs put the wind up traditional Swiss watchmakers, who feared that the product could take over and make analogue watches obsolete. They clubbed together to fund research into the use of quartz in watches, which they believed could ensure unrivalled accuracy. While they caught up with Bulova, they reckoned quartz wouldn't last and was merely a marketing gimmick.

Over in Japan, Seiko was also diligently researching timekeeping, and in 1969, the company presented the public with the first quartz watch. It was a disaster; riddled with technical faults, a uniquely ugly design and priced at an unreasonable US$1,250, the model was recalled when barely 100 watches had been sold. However, Seiko had opened the door on quartz design, and its assets proved too alluring for even the Swiss to resist, though they left digital technology to the American and Japanese companies.

The next phase in the history of the digital watch was the digital light emitting diode (LED) display. Hamilton was asked to design a futuristic clock for Stanley Kubrick's 1968 sci-fi film *2001: A Space Odyssey*, which gave the company's work an inspirational boost. Although competition intensified throughout the 1970s, digital watches were still too expensive for most, and no one in the watch industry was ready to abandon the analogue quartz watch just yet. Texas Instruments, also involved in manufacturing calculators, produced the first affordable LED watch in 1975, priced at US$20.

The last significant development in the saga of the digital watch was the liquid crystal display (LCD). Once again, the trail was blazed by Seiko and pretty soon everyone was following suit. In the 1980s, various additions were tested out on the wristwatch, such as TV screens and radio receivers, but in general these failed to catch on. Today, with the popularity of mobile phones comprising digital clocks, the very purpose of the digital watch seems uncertain. The analogue pocket and wrist watches will endure due to their value as symbols of wealth, taste and status, as well as their aesthetic appeal, but it appears that the inexpensive digital watch, as futuristic as it once seemed, may go the way of the dinosaur.

DINNER JACKET

When a male formal diner dons a dinner jacket or tuxedo, he inadvertently pays homage to an Algonquin chief called P'tauk-Seet. The chief lent his name to a region he once ruled about 40 miles north of New York City. In the seventeenth century, P'tauk-Seet – which is variously translated as 'bear', 'wolf' and 'clear running water' – was anglicised to 'tuxedo'. Later, a settlement called Tuxedo Park sprang up in the area.

In 1885, a resident of Tuxedo Park named James Brown Potter was introduced to the then Prince of Wales (later Edward VII) during a trip to Britain. The prince had recently taken to wearing a jacket that resembled a tail coat without the tails. According to legend, he first wore it when he became overwhelmed by the heat during a trip to India and ordered a tailor to dock the tails of his evening wear. A more likely explanation is that the garment was a 'Cowes jacket', which was worn by keen yachtsmen like the prince as a compromise between formality and practicality. At any rate, Potter was impressed and brought the idea back to his home town. The question was which member of the etiquette-obsessed upper echelons of the community would be daring enough to give the new form of attire its first public airing. Pierre Lorillard IV, who was heir to a tobacco fortune and the richest man in town, had several dinner jackets made up for the 1886 Tuxedo Park Autumn Ball, but got cold feet at the last moment. His son Griswold had no such scruples. He and his friends proudly sported the jackets at the ball. The residents of Tuxedo Park were too in awe of the Lorillards to be scandalised and the fashion spread like wildfire.

James Bond shows us how to wear a DJ.

DISCOTHÈQUE

Various establishments have claimed the title of the world's first discothèque, including Chez Regine, which opened in Paris in 1957, and Whisky A Go Go in Los Angeles, which was started in 1964. In fact, the first person to think of playing records non-stop to a crowd of revellers was an eccentric Englishman named Jimmy Saville (b1926). The art of DJ'ing made its debut in 1943 in an unlikely venue, the upstairs room of a working men's club in Otley, West Yorkshire. Saville later took to using two turntables to ensure that the music never stopped. He later became famous for his charitable work on behalf of spinal injury sufferers and as the presenter of a BBC television show called *Jim'll Fix It,* in which he arranged for the dreams of viewers to come true.

DOLLS' HOUSE

Albert V, Duke of Bavaria commissioned the earliest known dolls' house, between 1550 and 1579. It mimicked his own home exactly, and was affectionately referred to as his 'baby house'. A variation or offshoot of this was the fashion for merchants to display small, precious items in cabinets, which became known as 'cabinet houses'. Dolls' houses became popular among the upper classes in England, Holland and Germany in the seventeenth and eighteenth centuries, but were enormously expensive, as a fully furnished house required extensive labour by craftsmen. In England, the owners often wanted the dolls' house to resemble their own, inside and out down to the smallest detail.

There are several famous dolls' houses in existence, the most noteworthy being Queen Mary's (1867-1953) dolls' house at Windsor Castle in England, which took three years to complete. Built to 1:12 scale, and worked on by some 1,500 craftsmen and artists, the project was overseen by renowned architect Sir Edwin Lutyens (1869-1944), who, at the outset of the project said: 'Let us devise and design for all time something which will enable future generations to see how a king and queen lived in the twentieth century, and what authors, artists and craftsmen of note there were, during their reign.' Significant miniature contributions to the dolls' house include Cartier clocks, Veuve Clicquot Champagne, Rolls Royce cars and a short story (500 words) by Sir Arthur Conan Doyle for the library.

The other most famous 'miniaturist' is probably Mrs James Ward Thorn, a widely travelled American collector who built up a substantial array of miniature items in the 1930s. Eventually she hired a studio and decided to create custom-built interiors (more rooms than houses) in which to display the collection. A selection of her rooms were offered to the Art Institute of Chicago in 1940, who promptly sold them to IBM. Years later, Mrs Thorn's son spotted the rooms in a store window and arranged to have them returned to her for restoration. They are still on display in the Thorn Rooms at the Art Institute of Chicago.

Before World War I, Germany was the most prolific manufacturer of dolls' houses, but by the late 1940s there were factories throughout Europe and North America and many miniature items were in mass production, allowing it to become a popular hobby taken up by many enthusiasts.

DRAWING PIN

The drawing pin was invented in 1903 by a German clockmaker named Johann Kirsten. Hitherto, people wishing to pin up documents had been forced to use nails or map pins, which were invented in 1900. Kirsten sold the rights to his flat-headed pin to a businessman called Otto Lindstedt, who patented it on 8 January 1904. Lindstedt made a fortune from the simple device, which is known as a *pinne* in Germany. He is commemorated in a brass plaque in his home town of Lychen.

DRIVE-IN RESTAURANT

For some reason, establishments named in honour of the pig loom large in the history of innovation (see Supermarket). The first drive-in restaurant was a place called the Pig Stand, which opened for business in Chalk Hill Road, Oak Cliff, Dallas in 1921. The man with the plan was a Texan entrepreneur called Jesse G Kirby, who observed that, 'people with cars are so crazy they don't want to get out of them to eat'. To open the first Pig Stand, he teamed up with a local doctor named Ruben W Jackson. When customers pulled up to the establishment, they were greeted by uniformed youths who took their order, ran inside to collect the food and handed it through the car window on their return.

A drive-in waitress delivers food to a waiting car.

The Pig Stand chain expanded to over 100 locations, but Kirby hadn't finished yet. In 1931, he introduced a drive-through variation at a branch in San Antonio, Texas, setting the precedent for the 'drive-thru' restaurants of today.

E

US Patent 188,292 13 March 1877

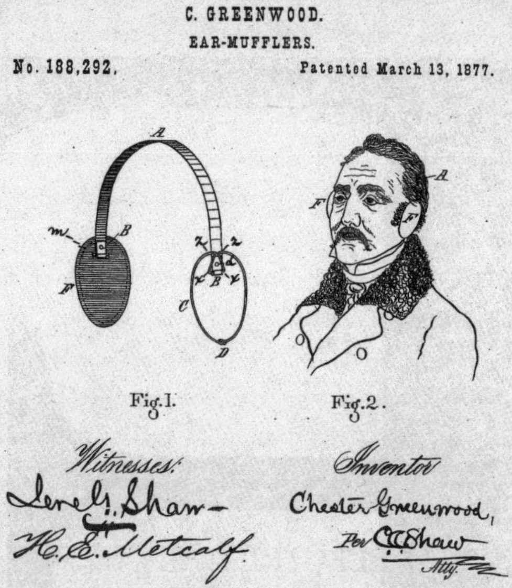

George Greenwood's earmuffs patent.

EARMUFFS

The story of earmuffs began with a 15-year-old boy living in Maine who had big ears and extreme sensitivity to the cold. Chester Greenwood (1858-1937) went to go ice-skating one day, but was driven indoors by the cold, which made his ears numb. He was allergic to wool, so could not wear the mufflers most people did. In desperation he fashioned together the rudimentary earmuff with help from his grandmother. The design and idea could not have been more simple; essentially a piece of wire, with some soft insulating material attached for each ear, and a few stitches. He opted for a combination of beaver fur on the outside and velvet on the inside. His invention caught on and other children in the area were soon pestering their parents to make them a pair. Greenwood was not entirely satisfied with his design though and tried out a flat, spring steel for the band, which enabled him to attach hinges to the muffs for a snugger fit. It also meant that he could coil it flat and put it in his pocket. Their fame began to spread and soon everyone wanted a pair of Greenwood's furry ear protectors.

On 13 March 1877, the United States Patent and Trademark Office awarded Greenwood patent No 188,292. At just 18 years old, he was sitting on a goldmine. He opened a factory down the road from his home, but before long he had to move premises, as the market was expanding so much. In 1883, his factory was producing 30,000 pairs of earmuffs a year, and by 1936, the annual output had risen to 400,000. He did particularly well during World War I, selling earmuffs to the army.

By the time he died in 1937, Greenwood was a Maine celebrity. His designs created an employment boom in the area and 21 December was even declared Chester Greenwood day in his honour. It is not known whether everyone wore earmuffs for the day to mark the occasion.

ELECTRIC GUITAR

The electric guitar was inspired by a 50:50 combination of physical laws and Hawaii. LA-based guitarist George Beauchamp (1899-1941) – who played Hawaiian music in a band – wanted to adapt the guitar in order to enhance its sound. Working at his kitchen table, he refitted a guitar body to sport two horseshoe magnets, with the strings running between them. Wrapping the strings in a magnetic field would enable their sounds to be picked up electrically and amplified, giving it just the kick he needed for his big band sound. Assisting him in the making of the 'Frying Pan', as the first electric guitar was named, was one Adolph Rickenbacher (1886-1939), who later changed the spelling of his name to Rickenbacker. However, there was a six-year lag between the invention of the Frying Pan in 1931 and its receipt of a patent, and in the interval other creators had set to work.

Among these was Les Paul (b1915). A jazz and blues musician, Paul had become frustrated with the new guitar's lack of versatility throughout the 1930s and started making his own. Realising that electrical signals meant the instrument no longer needed a hollow body, he increased the guitar's solidity with his innovation nicknamed 'The Log', which consisted of a fence post attached to the body of an Epiphone jazz guitar. Though it failed to impress the top brass at the Gibson Guitar Corp when Paul first approached them, he had nonetheless made a breakthrough. Gibson and Paul eventually went into partnership on mass-producing solid guitars in the early 1950s: just in time for rock 'n' roll.

Running alongside Paul's achievements were developments made by Leo Fender (1909-1991), helped by his musically-minded friend, Doc Kaufman. As Gibson's Les Paul took off, so too did the first Fender

US Patent 2,089,171 10 August 1937

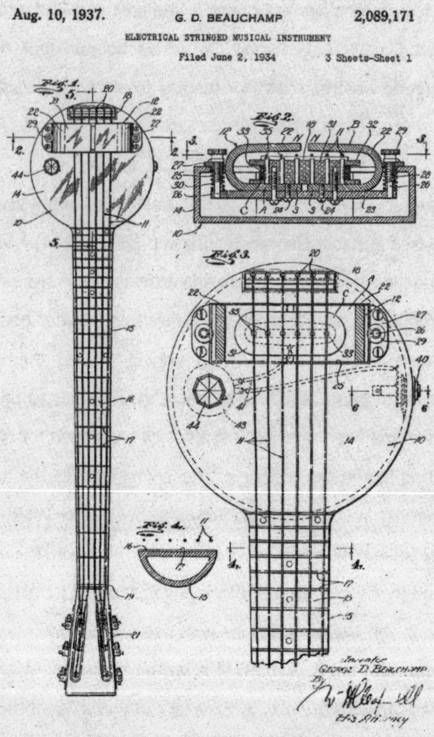

George Beauchamp's 'frying pan' – the first electric guitar.

model, the Esquire, which was quickly followed by the Broadcaster and Telecaster. By the mid-1950s, parents of the first teenagers must have been cursing the likes of Beauchamp, Paul and Fender as their offspring sought to emulate the heroes of the new rock 'n' roll age, probably long before they knew how to tune up.

In the 1960s, a new generation of musicians emerged in Britain in the wake of bands like The Beatles and Rolling Stones, who had taken American blues in more experimental directions. As the 1960s went on, groups continued to try out new sounds. Kinks guitarist Dave Davies (b1947) achieved the influential growling riff of classic track 'You Really Got Me' by cutting the cone of a 10-watt speaker with a razor blade. The increased vibrations resulted in greater distortion, and it can be argued that these simple razor-cuts gave birth to heavy metal.

Introduced in 1954, the Fender Stratocaster has become the world's best-known electric guitar. To this day, all the coolest musicians still play it.

ELECTRIC RAZOR

First patented by Colonel Jacob Schick (1878-1937) in 1928, the prototype of the electric razor was nothing like what we see on the market today. A bit cumbersome by our standards, the machine required a small external motor to power the shaving head, which needed to be 'kick-started' with an exposed turn wheel. The American public did not initially share the inventor's enthusiasm for the waterless wonder.

The electric razor was not Schick's first invention, nor did it show as much initial promise in the marketplace as his first innovation of 1921 had; a safety razor with replacement blades ready to be reloaded at the stroke of a built-in lever. Having developed a product that would prove the forerunner of today's popular injector razor, Schick could have simply stopped there. Yet his determination to provide the world with an electric dry shaver was a dream that would not die easily.

Schick had first conceived of his dry shaver during one of the many cold winters he spent staking mining claims in Alaska and British Columbia following his retirement from the US Army. On one of these trips, a sprained ankle made him a prisoner of his own camp. Schick managed to exist in -40°F temperatures by killing a moose for food. However, the process of shaving with freezing water was less than pleasant, and inspired him to come up with a drier, more convenient method. Convinced he had struck gold with the idea, he put all of his faith and money into the invention, even mortgaging his own home for $10,000. Schick went to work on improving his original design. Business grew steadily and by 1937, 1.5 million of his innovative razors were in use across the globe.

ELECTRIC TOOTHBRUSH

Dr Scott's electric toothbrush was a popular gadget in the 1880s, but it wasn't really an electric toothbrush at all; it was a simply a standard toothbrush with a magnet inside the handle, which Dr Scott thought gave him the right to say that it was 'permanently charged with electro-magnetic current'. He also created the 'electric' hairbrush and flesh-brush (whatever that is...), and even made a brush for horses.

Scott boasted that his brushes' medical benefits included cures for everything from headaches and blood diseases to constipation. 'There need not be a sick person in America (save from accidents), if our appliances become a part of the wardrobe of every lady and gentleman, as also of infants and children,' he puffed. It seems that the doctor's primary skill lay in marketing as he cleverly patented the ornate handles. This allowed him to call his product the 'patented electric toothbrush', which convinced people that it was patented for health benefits. It is not known whether Scott believed his own claims.

The first real, oscillating electric toothbrush emerged in 1939 from the workshop of Swiss inventor, Dr Philippe Woog. It had bristles of nylon, which had been invented the previous year, a happy improvement on hog's hair. It wasn't a hit in America until 1960; even regular, non-electric toothbrushes were relatively new in the US and only widely used there after World War II, and to ask the customer to put into their mouths a toothbrush that plugged into a wall socket – as the first version required – was perhaps a step too far. In the 1960s, a revised version was marketed by Squibb Pharmaceutical under the brand name, Broxodent. General Electric brought out the first cordless, rechargeable toothbrush the following year, which seemed to do the trick; since then sales have been consistently strong.

A further development took place in 1983, with the invention of the 'sonic toothbrush'. Delivering over 30,000 strokes per minute – compared with between 2,500 and 7,500 for the electric, and around 300 for the elbow-powered standard – the high speeds cause the bristles to vibrate and clean more effectively. It also stimulates saliva, the body's own anti-plaque mouthwash, energising its natural defences and cleaning the parts that other toothbrushes cannot reach. Philips was the first company to market the sonic toothbrush and called it 'Sonicare'.

ELEVATOR

The most primitive form of elevator is the hoist, which raises objects by means of man or animal power in combination with a system of ropes and pulleys. Archimedes (287-212BC) improved the design of the early hoist in the third century BC by adding a revolving drum or capstan for the rope to coil around as it was pulled in. Similar technology was in use at the Coliseum in Rome by 80AD to transport gladiators and animals from the basement to ground level. Louis XV (1710-1774) had a manually operated elevator installed in his apartment in Versailles in 1743.

Mechanical elevators powered by steam or hydraulics emerged during the first half of the nineteenth century, but their use was confined to industry. People needed to be utterly convinced that 'moving rooms' were safe before the passenger elevator could become a reality. The man who provided the necessary reassurance was the American Elisha Graves Otis (1811-1861). He devised a spring-loaded mechanism that automatically arrested the fall of an elevator if its hoisting rope snapped. In such an event, brakes would pop out of the car and engage with racks at the sides of the shaft. Otis exhibited his

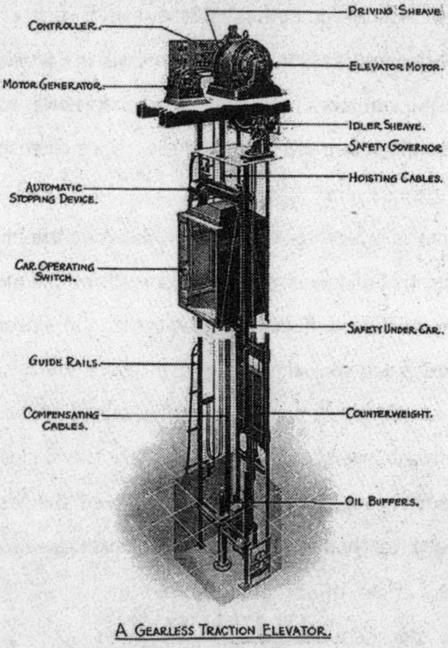

Diagram of a 1920s elevator.

invention at the Crystal Palace Exposition in New York in 1853, and was so confident of its efficacy that he repeatedly had the rope cut while he stood in the elevator car. It came to a rapid halt every time. On 23 March 1857, Otis installed the world's first commercial passenger elevator in the EV Haughwout & Company department store in New York City.

ENGAGEMENT RING

The practice of wearing engagement rings may have been initiated by the Ancient Romans. They wore them on their right hands as they considered the left side of the body 'sinister'. This remained the norm in the Christian world until the sixteenth century; people in some Christian European countries today still wear them on their right hands. The Roman betrothal ring or *anulus pronubus* was often made of iron. The romantic interpretation is that this was due to the strength of the metal. A more cynical view holds that iron was used to symbolise the fetters worn by slaves.

Engagement rings were endorsed by two popes. In 860AD, Pope Nicholas I (c820-867AD) instituted a betrothal ceremony in which the future groom was required to present his bride-to-be with a golden ring, perhaps to demonstrate his solvency. At the Fourth Lateran Council, 355 years later, Pope Innocent III (c1161-1216) decreed that there should be a long gap between betrothal and marriage. He advocated the use of an engagement ring to prove serious intent on the part of the man and to advertise the fact that the woman was spoken for.

ESCALATOR

The forerunner of the modern escalator was an inclined conveyor belt patented by Jesse W Reno in March 1892, which carried passengers uphill at an angle of 25 degrees. The invention caused so much excitement that Reno installed a version in Coney Island, where it became one of the amusement park's most popular rides. In the first few months, some 75,000 people took the opportunity to be raised about seven feet. In 1898, a Reno Inclined Elevator was built in Harrods in London. A porter was stationed at the top to hand out brandy to any customers overwhelmed by the experience.

The other great escalator pioneer was Charles D Seeberger, who joined forces with the Otis Elevator Company to produce the first practical device with moving stairs. The invention was exhibited at the Paris Exposition of 1900, and a commercial version was installed at Gimbel's department store in Philadelphia the following year. It was Seeberger who coined the term 'escalator', which he arrived at by eliding 'elevator' with *scala,* the Italian word for 'stairs'.

ETCH A SKETCH®

During the late 1950s, a Frenchman named André Cassagnes invented a remarkable toy. It had a glass screen coated on the inside with

US Patent 3,055,113 25 September 1962

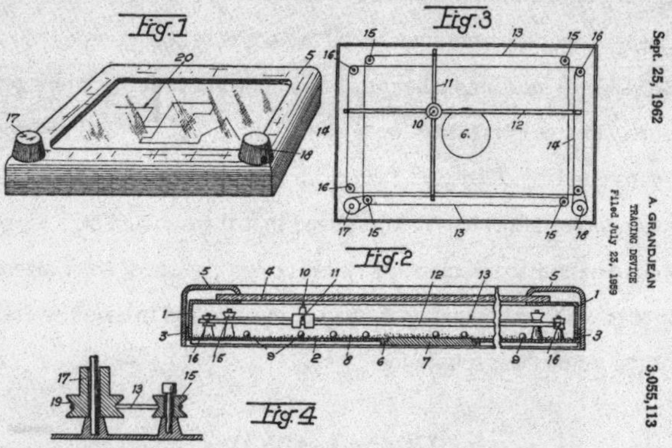

Cassagnes' 'Magic Screen' or Etch A Sketch.

aluminium powder and tiny styrene beads and a stylus that could be moved vertically or horizontally via a pair of knobs. As the stylus moved, it scraped away some of the beads leaving a distinct black line. To restore the screen to its original pristine condition, all that was needed was to give the toy a good shake. Cassagnes decided to call his invention *L'Ecran Magique* ('The Magic Screen'). He took it to the Nuremberg Toy Fair in 1959 and licensed it to the Ohio Art Company. The first Etch A Sketches, as the toy was renamed, were manufactured on 12 July 1960. It may be virtually impossible to use it to draw curved lines, but Etch A Sketch is still one of the world's most popular toys with over 100 million units sold to date.

F

FAX MACHINE

If you ask a random sample of people to guess when the first fax was sent, they usually plump for some time around the 1960s. The remarkable truth is that the fax machine was born in 1843. The apparatus was assembled by a Scotsman named Alexander Bain (1818-1903), and consisted of two pens connected to two pendulums joined by a wire. The primitive device worked via a scanning detector, which sent out electrical signals of differing strengths depending on whether it was passing over the black areas of an image or the white ones. The machine was able to reproduce writing on chemically treated paper by means of a receiver at the other end of the wire.

FERRIS WHEEL

The first Ferris wheel was built by a man named, unsurprisingly, Ferris. George Washington Gale Ferris (1859-1896) was a bridge builder from the 'steel city' of Pittsburgh, Pennsylvania. In 1892, he was commissioned to build an attraction to rival the Eiffel Tower for the World's Fair in Chicago the following year. Ferris came up with a monstrous wheel 250 feet high,

Ferris's giant wheel at the Chicago World's Fair.

supported by two 140-foot-high towers connected by a 45-foot axel, the largest single piece of steel manufactured to that date. The ride had 36 cars, each holding 60 passengers who paid 50¢ for the privilege. It was powered by two 1,000-horsepower engines. The wheel was later dismantled and moved to Missouri for the 1904 St Louis Exposition. It was decommissioned two years later.

FIRE

It is likely that there were two stages in our ancestors' use of fire. First they learned how to maintain it, then they learned how to make it. Some archaeologists believe that the former, which entailed transporting and tending fires generated by lightning or volcanic activity, may have happened as early as 1.5 million BC. The analysis of scorched animal bones of this age found at Swartkrans, north of Johannesburg in South Africa, suggests that they were heated to temperatures only usually reached in deliberately constructed hearths. There is less controversy about a group of bones and stone tools found at Zhoukoudian in China. They are generally taken to prove that *Homo erectus* was building and maintaining fires some 400,000 years ago. The first hominids to make fires from scratch are thought to have been the Neanderthals.

FISH AND CHIPS

Deep fried fish and fried potatoes were eaten separately for a long time before anyone thought of combining them in one meal. Sephardic Jewish immigrants introduced the British to the delights of fried fish during the seventeenth and eighteenth centuries. By the middle of the following century, the dish had become a staple of working class Londoners. Charles Dickens mentions a 'fried fish warehouse' in his novel *Oliver Twist*, which was serialised in 1838. Around the same time, 'chippies' became popular in the North of England, probably as a result of 'potato shops' opened by Irish labourers. The first person known to have united the two trends was an Eastern European immigrant named Joseph Malin, who opened a fish and chip shop in Cleveland Street in London in 1860. The British took to the combination in such a big way that it soon became regarded as the national dish.

FOOTBALL

The practice of kicking a ball around has existed for thousands of years in varying forms and in many different cultures around the world. The oldest recorded game, which bears a strong resemblance to the modern version, occurred in China in the second and third centuries BC. A military manual mentions the game of 'Tsu'Chu' as a training exercise. This involved trying to kick a ball measuring 30-40cm filled with feathers and air and into a small net fixed on to bamboo canes. Other versions mention the kicker having to fend off attackers to hit the target. Like modern football, the use of the hands was strictly forbidden. The Romans also played a game reminiscent of what we play today. Played on a pitch marked out with boundary lines and a central line, the object of the game was to get the ball over the opponents' boundary line. Underhand tactics and manoeuvres were vociferously encouraged by the enthusiastic crowd.

The Romans would have taken this game to Britain with them, but it is uncertain whether it formed the basis of today's game. Between the eighth and nineteenth centuries in Britain, it evolved into many different regional styles and versions. One thing these versions did share with the Roman game was their uncivilised nature. In many areas, it was acceptable to kick your opponent and generally obtain the ball by any means possible. One form of the game being played was 'mob football' where the number of players was unlimited and the rules were vague, meaning it was quite a riotous affair.

Although the precise details of how this game evolved into what we play today remain unclear, it was certainly a popular pastime, much to the dismay of the authorities. Throughout the thirteenth and fourteenth centuries, it was viewed as public nuisance and uncouth. It was particularly frowned upon in the sixteenth century, when the rise of Puritanism labelled it as frivolous and prohibited playing the game on Sundays.

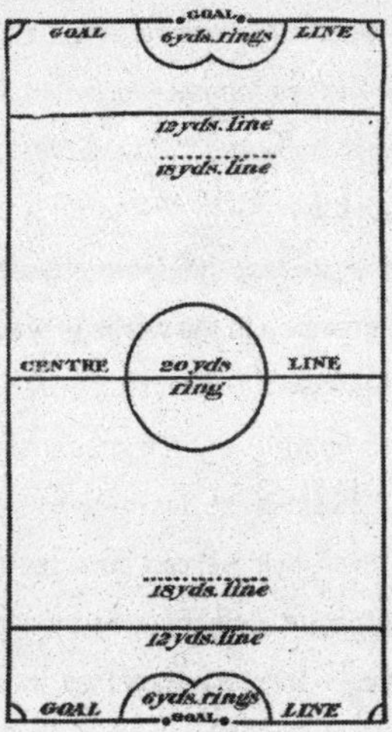

The field in 1892

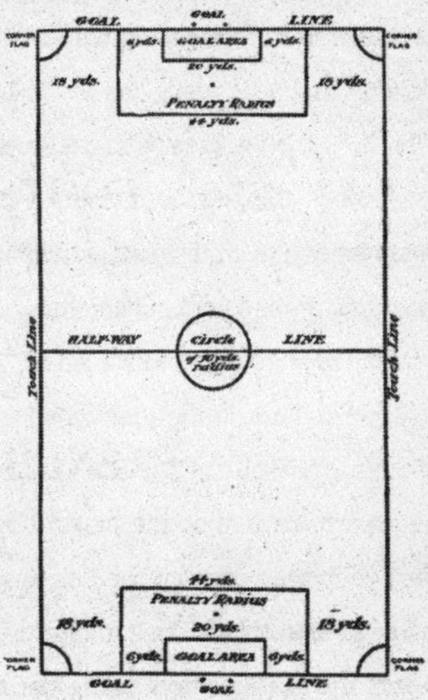

The perfected playing pitch in 1905.

The evolution of the soccer pitch.

It was not until the nineteenth century that the history of football reached a turning point, when school football was introduced. Each school played its own particular style and variations on the rules, but as the educational benefits of playing became apparent, schools sought to try and standardise the rules. A notable division occurred between the schools, which ultimately saw the birth of two different games; football and rugby. At Rugby school in 1846, the first standardised rules of the game were laid down, where carrying the ball was allowed and even kicking one's opponent in the shins was deemed acceptable. Many other schools adopted Rugby's rules, but others such as Eton, Harrow and Winchester rejected carrying the ball, preferring to kick it.

Two distinct versions of the game therefore began to evolve, but it was not until 1863 that the two games were formalised. Intent on sorting out the confusion, Cambridge University organised a meeting of 11 London clubs and schools in a London pub where they discussed the rules, and The

Football Association (FA) was officially born. While most members agreed that carrying the ball should not be allowed, Rugby stood firm to its rules and later that year, rugby and football went their separate ways.

By 1871 the FA already had 50 member clubs and this same year the first football competition in the world took place – the FA Cup. Football spread slowly to other parts of the world, but by 1907, New Zealand, Argentina, Switzerland, Chile, Italy, Germany, Uruguay, Hungary and Finland had all founded their own associations.

FRENCH FRIES

The French insist that the practice of deep-frying sliced potatoes was invented in their country. The Belgians are equally adamant that the dish was created in theirs. The controversy may never be resolved, so we will present the evidence from both sides and let the reader decide.

The French claim of precedence is supported by no less a person than Thomas Jefferson (1743-1826), the third president of the United States. The first unambiguous reference to what are now known as French fries occurs in a manuscript written by Jefferson around 1801. He describes 'potatoes deep-fried while raw, in small slices'. This was a daring concept at the time as potatoes were considered highly poisonous unless they had been thoroughly boiled. Nevertheless, Jefferson undoubtedly served 'potatoes fried in the French manner' to guests at Monticello, his mansion in Virginia. This is confirmed by a surviving printed menu. He also had a French chef called Honoré Julien.

The Belgian case rests on Jo Gerard, a respected historian who claims to have proof that his countrymen were eating 'French fries' as early as the seventeenth century. Gerard bases his claim on an unpublished document dating from 1680. He says it describes contemporary residents of the area between the cities of Dinant and Liège in the Meuse valley eating deep fried potatoes in the depths of winter. The poorer inhabitants of the region were in the habit of eating small fish fried in oil. When the rivers froze, they apparently cut potatoes lengthways and fried them instead.

FRISBEE®

Ever wondered where the idea of throwing a saucer-like piece of plastic through the air came from? Combine a group of students and some pies and you might have an answer. A baker named William Russell Frisbie of Connecticut, came up with the ploy of putting his family name in relief on the bottom of the tin pans in which his pies were made. The pies became a staple diet for students at many New England colleges, and – never short on inspiration for making their own entertainment – a group of students soon discovered that the empty pie tins could be tossed and caught, providing hours of fun. The issue of who was the first to fling is the source of much debate, but the commercial story is taken over by Walter Frederick Morrison.

In 1948, Morrison and his partner Warren Franscioni designed a plastic version of the flying pie tin, which, unsurprisingly, had much better aerodynamics than a tin plate. The Morrison/Franscioni partnership did not last long, but Morrison went on to produce another version called the

US Patent 183,626 30 September 1958

United States Patent Office

Des. 183,626
Patented Sept. 30, 1958

183,626
FLYING TOY
Walter Frederick Morrison, La Puente, Calif.
Application July 22, 1957, Serial No. 47,035
Term of patent 7 years
(Cl. D34—15)

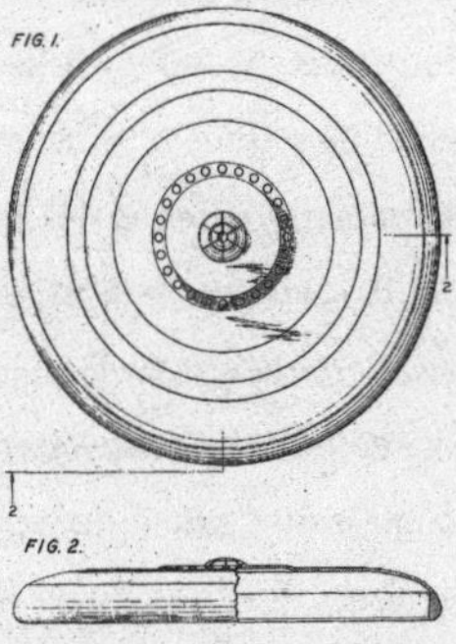

Walter Morrison's original flying disc patent.

Pluto Platter, capitalising on the American fascination with UFOs and flying saucers. The Pluto Platter became the blueprint Frisbee and the outer ring became known as the Morrison slope.

The invention caught the eye of Richard Knerr and Arthur 'Spud' Melin at their new toy company Wham-O', promoters of the Hula-Hoop (see Hula-Hoop) and the SuperBall. Morrison was persuaded to sell the rights to his design and Wham-O began production of the Pluto Platter in 1957. While looking for a catchy new name to boost sales, Richard Knerr heard the story of the Frisbie pies, and he changed the spelling to create the registered trademark 'Frisbee'. By promoting it as a new sport in the college tradition, Knerr's re-branding paid off and sales soared. Today Frisbee has a massive cult following. For the more serious and competitive, there is the world of Ultimate Frisbee, invented by high school students in Maplewood New Jersey in 1967.

FROZEN FOOD

The ability of ice to preserve food has been known since ancient times. The Mesopotamians were constructing ice-houses using snow gathered from mountainous regions as early as 1,700BC, and the Egyptians were not far behind them. Meanwhile, the residents of the Andes have been freeze-drying potatoes since prehistoric times, crushing them and spreading them on rocks at altitudes in excess of 15,000 feet to make a product called *chuno*. But it was the Inuit, whose food freezes solid for much of the year whether they want it to or not, who provided the stimulus that led to the modern freezer. In 1912, a New Yorker named Clarence Birdseye (1886-1956) went to work as a fur trader in icy Labrador to fund his education. He watched the local Inuit freezing the fish they caught, and noticed that the temperature at the time made a significant difference to the eating quality once they were defrosted. The fish tasted particularly fresh if they had been frozen at temperatures below -40°C/-40°F. Birdseye reasoned that this was because the process happened so quickly that large ice crystals were

US Patent 1,773,079 12 August 1930

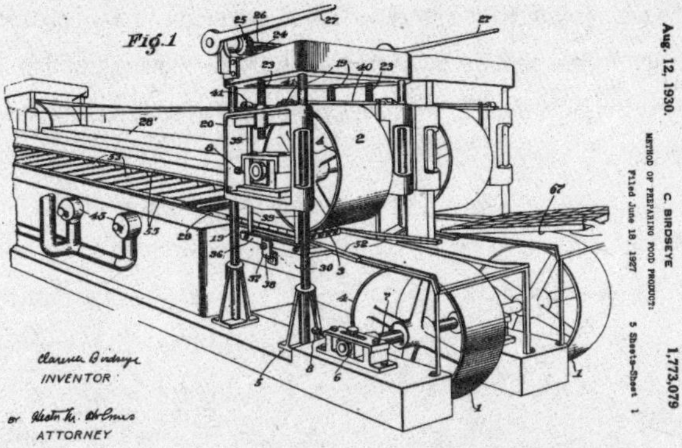

The original patent drawing for Clarence Birdseye's Quick Freeze Machine, which comprised a refrigerating space and a conveyor belt for transporting the foods through the process.

unable to form in the flesh, which would otherwise have been damaged by them.

When he returned to the US, Birdseye spent several years working on a way to duplicate Arctic conditions in the factory or home. In the mid 1920s, he unveiled his patented 'Quick Freeze Machine', which reduced the time needed to freeze food from three days to a matter of minutes.

FUNERAL

The ritual burial of the dead is often regarded as uniquely human, but astonishingly, the practice may well have been invented by an entirely different species. The oldest archaeological evidence suggests that the first funerals were carried out by our extinct cousins the Neanderthals. While conducting excavations in Shanidar Cave in the Baradost Mountains of Iraq during the early 1960s, Rose and Ralph Solecki (b1917) discovered what appeared to be the 60,000-year-old grave of an adult male Neanderthal. Pollen deposits on the bones indicated that the presence of garlands of wild flowers and medicinal herbs. It was difficult to avoid the conclusion that he had been buried deliberately and with reverence.

The discovery of Neanderthal burial sites has momentous implications, although in the absence of written records they remain just that. The evidence suggests that *Homo neanderthalis* may have had a conception of an afterlife.

G

GARDEN HOSE

An old rubber hose factory in Delaware.

The first hoses were made of ox gut in about 400BC. They were operated by stamping on water-filled bags attached to the end of a hose, which caused water to squirt out of the other end. There was little innovation over the next 2,000 years, but in 1673, Jan and Nicolaas van der Heisen developed a hose made of stitched leather for the Dutch fire service. Although leather hoses were wont to burst, they monopolised the field until 1871, when the BF Goodrich Company of Akron, Ohio introduced a rubber version reinforced with cotton. This hose, one of the earliest commercial applications of rubber, immediately became the company's best seller. The first customers were fire departments, but gardeners soon got wind of the new technology. By the 1880s, the watering can had been relegated to a secondary position in the horticultural toolkit.

GOLF

The Scottish, who pride themselves as the inventors of golf, won't thank us for saying so, but the sport probably evolved in the Low Countries such as the Netherlands and Belgium. The direct ancestor of golf appears

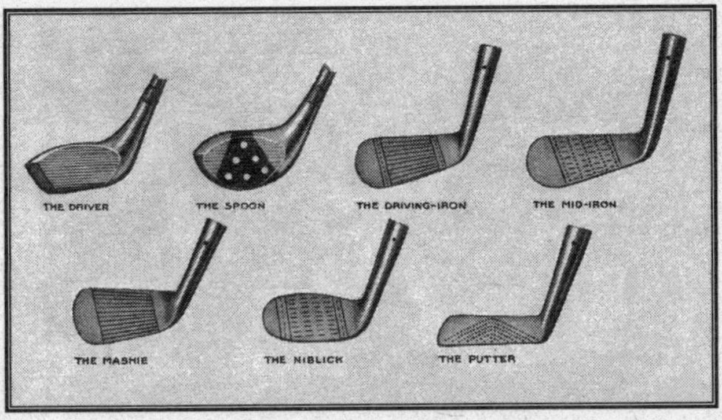

to have been a game played in France and Belgium called *chole*, which is mentioned in a document dating from 1353. *Chole* certainly involved using sticks with curved bases to hit balls from fixed starting points to fixed finishing points, but there were no holes. There was a good reason for this: *chole* was played on ice. It was only when the game reached Scotland that holes were introduced, for the simple reason that the locals wanted to play throughout the year. It is thought that *chole* was brought to Scotland by Hugh Kennedy, Robert Stewart and John Smale in 1421.

The first reference to golf as such occurs in a decree banning the sport issued by King James II's Scottish parliament in 1457. The game was thought to be interfering with archery practice. For good measure, James also forbade the playing of football.

GREENHOUSE

Surprisingly enough, the greenhouse didn't have to wait for the large-scale production of glass to come into existence. The Romans got there first. In 30AD, the Emperor Tiberius was advised by his doctor to eat one cucumber every day. To grow the vegetables out of season, the imperial gardeners hit upon the idea of using large pots covered by sheets of the translucent mineral mica.

The greenhouse at the Jardin d'Acclimatation in Paris.

The modern greenhouse came into being during the sixteenth century, when explorers began to return from far-off lands with exotic plants that were unable to survive European winters without help. The first practical glass-based greenhouse was built in Leiden, the Netherlands, in 1599, by a French botanist named Jules Charles. The French were particularly active greenhouse builders in the early days as a result of their enthusiasm for oranges. The orangery at the Palace of Versailles was a vast structure measuring 45 feet high, 42 feet wide, and more than 500 feet in length.

The first recorded greenhouse in America was built by a Boston merchant named Andrew Faneuil in or around 1737.

H

HAMBURGER

Children are often surprised to learn that hamburgers do not contain ham. In fact, the name derives from the German city of Hamburg. The origins of the dish, however, are further afield, in the steppes of central Asia. The Mongolian horsemen who terrorised Eastern Europe in the thirteenth century used to make cakes of raw minced beef or lamb and store them under their saddles. This practice had two advantages: it tenderised the meat and allowed them to eat on the hoof. When the Mongols invaded Moscow, they brought their delicacy with them. As the locals called the invaders 'Tartars', the dish became known as 'steak tartare'. The recipe spread to Russia's ports on the Baltic Sea, where it was picked up by sea merchants from Hamburg. Patties of spiced minced beef proved very popular with the city's residents, particularly those who couldn't afford steak. They were eaten cooked or raw according to preference.

Mongolian warriors introduced their raw beef and lamb cakes to Moscow when they invaded. From there they spread to Hamburg, and eventually the US and beyond.

In the nineteenth century, German immigrants brought the 'Hamburger steak' as it was now known to the US. Several American towns claim to be the birthplace of the hamburger in a sandwich, including Athens, Texas, Hamburg, New York, New Haven, Connecticut and Seymour, Wisconsin. There is more certainty about the bespoke burger bun. It was invented in 1916 by Brian Peppers, who later founded the White Castle chain.

HANDCUFFS

They may only be everyday items for cops, incompetent criminals and, of course, the somewhat kinky, but handcuffs play a vital role in the maintenance of law and order. They have been around in one form or another for millennia – in the fourth century BC the Greeks discovered chariots full of handcuffs among the possessions of a defeated Carthaginian army – but prior to the 1860s they worked on the principle of 'one size fits all'. This meant that particularly large prisoners could be impossible to restrain, while their thinner-wristed colleagues sometimes simply slipped their irons. The solution was a handcuff with an adjustable ratchet mechanism. The first such device was patented by the American WV Adams on 17 June 1862.

Antique handcuffs from the American Civil War. A key allowed adjustment for each prisoner's wrist, unlike early cuffs that were 'one size fits all'.

HAPPY HOUR

The phrase 'happy hour' supposedly emerged in the US Navy during the 1920s. It originally described a scheduled entertainment period onboard ship during which boxing and wrestling bouts took place. This was a valuable opportunity for sailors to let off the steam accumulated during long periods at sea. The expression gradually spread to the civilian community. During the Prohibition era, it served as a useful euphemism for drinking sessions ahead of dinner in public restaurants where the serving of alcohol was illegal.

HEADPHONES

The earliest headphone wearers were workers in the telephone and radio industries. The 'cans' differed from modern versions in that they relied on signals too weak to produce audible sounds through regular speakers. The first models marketed to the public were the DT 48 'dynamic' headphones developed by the German company Beyerdynamic. They went on sale in 1937 and are still in production. Stereo headphones were invented by a jazz musician from Milwaukee called John C Koss in 1958. He designed them as part of a package that included a mini-phonograph and speakers, but his headphones were the only component to attract commercial interest. They were unsophisticated, consisting of tiny cardboard covered loudspeakers attached to military headbands, but they sparked an industry for which millions of commuters can be grateful.

HELICOPTER

The world's first helicopter 'flight' took place on 24 August 1907. The machine in question, the Gyroplane No 1, lifted its passengers all of two inches off the ground. It was built and 'flown' by the French brothers Jacques and Louis Breguet (1880-1955) and was a cross-shaped device with a rotor made from biplane blades at each of its four corners. The Breguets' fellow countryman Paul Cornu did somewhat better in November of the same year. His flight lasted 20 seconds and reached a dizzying altitude of a yard. Over the next decades, inventors constructed all manner of helicopter-like aircraft with varying degrees of success. They included the autogyro, which was invented by the Spaniard Juan de la Cierva and first flown in January 1923. Autogyros were effectively standard airplanes fitted with huge horizontal blades. As they needed forward motion to stay airborne they couldn't hover or ascend or descent vertically, but they otherwise functioned as helicopters and were used by the British military for surveillance purposes.

The modern helicopter was developed by Igor Sikorsky (1889-1972), who fled Russia for America after the Bolshevik Revolution in 1917 and founded

US Patent 1,848,389 8 March 1932

March 8, 1932. I. SIKORSKY 1,848,389

AIRCRAFT, ESPECIALLY AIRCRAFT OF THE DIRECT LIFT AMPHIBIAN TYPE AND MEANS OF CONSTRUCTING AND OPERATING THE SAME

Original Filed Feb. 14, 1929 8 Sheets-Sheet 1

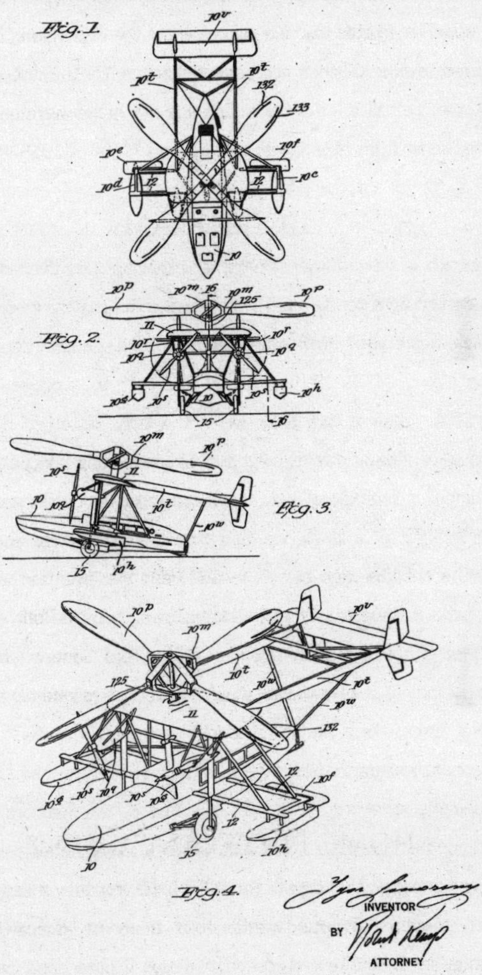

Igor Sikorsky's helicopter design, which formed the basis for the modern aircraft flown today.

the Sikorsky Aero Engineering Corporation in Long Island six years later. Sikorsky initially concentrated on building fixed-wing amphibious aircraft. This pattern continued after United Aircraft purchased his company, but he started to tinker with helicopter design on the side. In 1938, United was forced to close down its Sikorsky division as a result of the Depression, but the firm continued to finance Igor's helicopter researches. These culminated in the development of the VS-300, the world's first true helicopter. After months of tethered tests, its first free flights took place in 1939.

HOLOGRAM

A hologram is a two-dimensional image that appears three-dimensional under certain light conditions. The term, which means 'the whole message' in Greek, was coined in the late 1940s by a Hungarian scientist named Dennis Gabor (1900-1979). He did his best to produce convincing holographic images, but they were invariably distorted due to the technological limitations of the time. Gabor ruefully concluded that his 'experiment in serendipity' had 'begun too soon'. What was needed was a source of light of a single wavelength/colour emanating from a single point. The breakthrough came in 1960 with the invention of the laser beam (see Laser Beam). Two years later, Emmett Leith (1927-2005) and Juris Upatnieks of the University of Michigan read Gabor's original paper on holography. Out of curiosity, they decided to replicate his experiment using a laser. Their hunch produced the first convincingly three-dimensional holograms, which depicted a bird and a toy train.

HOME PREGNANCY TEST

An Egyptian papyrus written in about 1,350BC describes a pregnancy test which required the suspected mother-to-be to urinate on a patch of wheat and barley seeds over several consecutive days. If either seed sprouted, she was diagnosed as pregnant. She was held to be carrying a male child if the barley grew first and a female if it was preceded by the wheat. If neither seed germinated the test was considered negative. Research conducted in

1963 concluded that the method accurately diagnosed pregnancy 70% of the time.

The first modern home pregnancy kit was developed by the firm Warner-Chilcott. The EPT or 'early pregnancy test' was first marketed in the US in late 1977 and diagnosed pregnancy with 98% accuracy. Negative results were less reliable (80%). The outcome was revealed by thin blue lines on the instrument's display panel. Two lines indicated pregnancy; one meant that the test was negative. The anxious purchaser had to wait two hours for a reading.

HOROSCOPES

In its early days, astrology was the preserve of royalty who had their own personal astrologers. For the astrologer, this brought the advantages of a comfortable lifestyle in the royal palace, but it also had the disadvantage that they were often the deliverer of bad omens, over which they could quite literally lose their head. Horoscopes involve the practice of astrology, a complex process based on the position of the eight planets in the solar system, the sun and the moon and the time of a person's birth. These configurations are then used to forecast future scenarios.

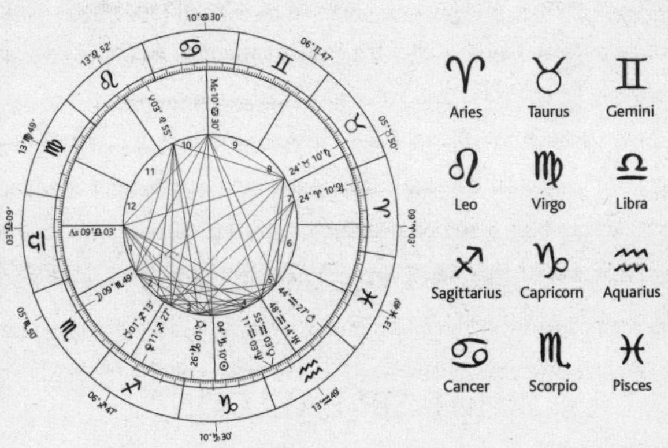

An example horoscope chart and, right, the 12 signs of the zodiac.

The study of astrology dates back to the beginning of civilisation and predates the origins of world religions. The Sumerians who lived in Mesopotamia around 4,000BC are the first example of people who worshipped the sun, moon and Venus. They considered them to be gods or the homes of gods. The priests who communicated with these gods were the first rulers and their purpose was to predict natural events such as weather and eclipses so they could maintain their power. While their practices were not astrology as we know it today, they did devise a calendar and identify the basic cycles of the sun, moon and stars.

Various civilisations through the ages built on this idea of using the sun and the moon to predict events, but it was the Greeks who had an immense influence on astronomy and astrology during the fourth and fifth centuries BC. The word 'horoscope' is actually Greek derived, meaning 'scope of the hour'. They were responsible for introducing mythology into astrology and fostering the zodiac and horoscope teachings still around today.

With the spread of Christianity in the Middle Ages, the practice of astrology became less popular and the Church in Rome procrastinated over its official standpoint. It was not until the sixteenth century, when the astrologer named William Lilly renamed it 'Christian astrology', that it became popular. Together with the Renaissance emphasis on science and discovery, the era saw a resurgence of astrology's popularity.

Today, opinion is still split over astrology, but it remains strongly engrained in popular culture. It is hard to find a magazine or newspaper that does not have a section dedicated to the astrological stars. It would seem that, even in the twenty-first century, many of us still enjoy the idea that our fate is written in the stars.

HOT-AIR BALLOON

The Chinese have been making paper balloons since at least the second century AD, but for some reason they never got round to constructing one large enough to carry humans aloft.

Montgolfier's first manned balloon flight soared to 6,000 feet.

Instead, the honour went to Joseph Montgolfier (1740-1810), the large and absent-minded twelfth child of a wealthy paper manufacturer from Viladon-les-Annonay in the South of France. There are several legends about how Montgolfier acquired the idea that would make him famous. They range from the sight of his wife's bloomers billowing with hot air as they dried over a fireplace to an incident in which a conical sugar loaf wrapper shot up the chimney when it ignited. The man himself, however, maintained that the moment of inspiration occurred one evening in 1782 when he was gazing at a print on the wall of his lodgings in Avignon. It depicted the ongoing siege of the island of Gibraltar, which the French and Spanish had been attempting to capture from the British since 1779. Montgolfier decided that there had to be some way to get troops onto the island. While entertaining this thought, he glanced at the fireplace and noticed sparks flying upwards. He gazed at the print again and ran out to buy some silk and sticks. When he returned to the apartment, he used them to construct a sphere with an aperture at one end, which he filled with twists of burning paper. The globe floated effortlessly to the ceiling.

Montgolfier wrote to his brother Étienne (1745-99) in a state of high excitement, promising him that if the latter brought him some taffeta and rope he would see 'one of the most astonishing sights in the world'. The brothers devoted the next few months to the development of the invention. This culminated in a public demonstration before the astonished citizens of Viladon-les-Annonay on 4 June 1783. The balloon, which had a circumference of 110 feet and was held together by 1,800 buttons, soared to an altitude of 6,000 feet within minutes. Barely four months later, on 21 November, a 'Montgolfier' launched in Paris bore the first human passengers skywards: they were Jean-François Pilâtre de Rozier (1754-1785) and the Marquis d'Arlandes (1742-1809).

HOT DOG

Any attempt to establish the identity of the first person to wrap a piece of bread around a sausage is, of course, futile. On the other hand, we are on firmer ground when it comes to pinpointing the originator of the term 'hot dog'. The main candidate, Harry Stevens, had a food concession at the polo ground, which was home to the New York Giants at the start of the twentieth century. One cold April day in 1902, Stevens was having trouble shifting ice-creams and sodas, so he instructed his vendors to get hold of some rolls and Wiener sausages. The Wiener was then colloquially known as the 'Dachshund', the German name for the hound that Americans call a 'Wiener dog'. To tempt the shivering crowd, the vendors shouted: 'Get your Dachshunds while they're red hot!' This cry was heard by 'Tad' Dorgan, the sports cartoonist for the *New York Evening Journal*. He had yet to come up with an idea for the next edition, so he decided to draw a roll filled with a frankfurter with the legs, tail and head of a Wiener dog. Being unable to spell 'Dachshund', Dorgan gave his cartoon the caption 'hot dog'. The name stuck, as did the tradition of serving hot dogs at baseball games.

Incidentally, there is a poplar but misleading rumour that John F Kennedy's famous statement of solidarity with the citizens of Berlin: *Ich bin ein Berliner* translates as: 'I am a hot dog.' Although there is a indeed a kind of sausage called a Berliner, literal-minded Germans are much more likely to have understood the remark as the President claiming to be a 'jelly doughnut'.

HOURGLASS

Nowadays sandglasses are only commonly used as egg timers, but when they were first invented they were invaluable nautical tools. Unlike mechanical

Heaving the log

At sea, a sandglass was used in combination with a knotted rope (or "log line") to measure sailing speed.

clocks, they were little affected by the movement and variations in temperature and humidity that were inevitable at sea. Hourglasses began appearing in ships' inventories around the year 1400. Small 28-second versions were used in combination with knotted rope to measure sailing speed, hence the term 'knots'. Larger, slower flowing examples were employed to keep time. They were used to divide the standard four-hour watch into half-hour segments.

Despite their nautical associations, the earliest evidence for the existence of hourglasses is found in a series of frescoes painted by Ambrogio Lorenzetti (c1290-1348) in the Sala della Pace of the Palazzo Pubblico in Siena, Italy, in 1338. One of the images depicts the allegorical figure of Temperance holding a large sandglass in her right hand. The hourglass would become a recurring motif in Western art used to remind viewers of the transience of human life.

HULA-HOOP®

It is likely that some form of hoop toy has been around since humans were able to wiggle their hips. Hoops constructed from vines or wood, used for exercise and in games, have been recorded in both ancient Egypt and Greece. In England in the fourteenth century hoops were also popular, though bizarrely 'hooping' was regarded as a fairly dangerous hobby which could induce back dislocations or heart attacks, and was eventually banned. The 'hula' part was added after British sailors visited Hawaii in the nineteenth century and noticed that the traditional dance, called the hula, contained very similar hip motions to the ones used to keep a hoop spinning around the waist. However, it was in Australia in the 1950s that the first hooping craze took place, mainly among school children who used bamboo hoops in dancing. By 1957 they were being properly manufactured, and were picked up in 1958 by the American company Wham-O, who were already producing the popular Frisbee. Arthur 'Spud' Melin and Richard Knerr, the founders of Wham-O, designed a compound called Grex (a form of polythene) specifically to manufacture their Hula-Hoops. By 1960, they had sold 100 million Hula-Hoops worldwide. But not everyone was as enthusiastic; in Japan the Hula-Hoop was banned, as the gyration of the hips was thought to be indecent; it was also banned in the Soviet Union where it was dubbed an example of the 'emptiness of American culture'. The Hula-Hoop is still available today in a plethora of colours and designs, though its popularity has never equalled the hysterical peak of the late 1950s.

I

ICE CREAM

There are two problems with the popular theory that Marco Polo (1254-1324) brought ice cream from China to Italy at the end of the thirteenth century, one silly and one serious. The frivolous objection is that it would have melted in his luggage. More persuasive is the argument that Polo failed to mention the dessert in his memoirs. On the other hand, there is little doubt that Italy was the first European country to experience ice cream, and that the Chinese were certainly pioneers in the field. During the early part of the Tang Dynasty (618-907AD), the Emperor employed 94 men to manufacture ice cream from buffalo and goat milk. It was thickened with ground rice and flavoured with camphor. The Chinese also discovered that mixing snow with saltpetre produced temperatures below freezing point and used the reaction to make the first ice cream machines. The earliest known ice cream-like concoction, however, was made in Persia around 400BC. It was a sorbet-like substance, the main ingredients of which were rosewater and vermicelli. It was frozen in giant yakhchals, 'refrigerators' cooled by ice gathered from mountain tops and served to royalty in the summer. The Iranians still eat a version of the dish, which they call *faludeh.* The Arabs were also key players in the history of ice cream, which was a favourite dish of the Caliphs of Baghdad. They were the first to add sugar and had built several ice cream factories by the tenth century.

ICE-SKATING

Ice-skating has a long history, and was first practised over 1,200 years ago by Vikings in Sweden. Those forerunners of today's skates were rather primitive, though. The original model was made from reindeer or oxen bones. In fact, the word 'skate' in Dutch is *schenkel* meaning 'leg bone'. The bones were filed down to make a smooth surface and were tied to the feet using leather straps.

Skates were developed further in the fourteenth century, when the Dutch attached iron runners to a wooden base. This was then fastened to the shoe using leather straps. They were far from efficient and

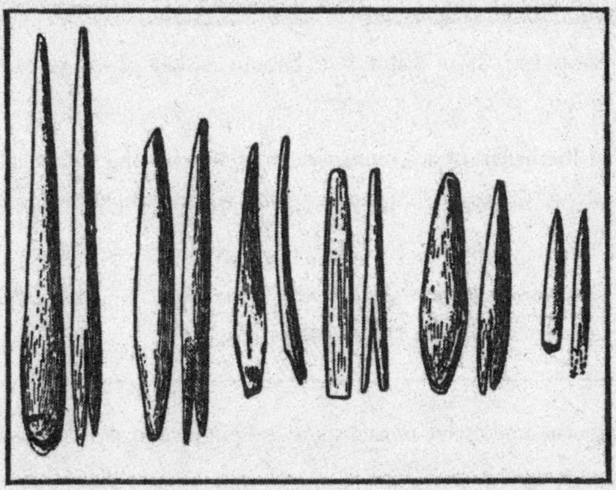

Reindeer and oxen bones, filed down and tied to the feet, were used by the Vikings as primitive early ice skates.

needed a pole to allow the wearer to push themselves forward without falling over. In cold countries, skates were used primarily as a form of transport over frozen lakes and rivers. By the fifteenth century, the Dutch had added a double-edged blade to the wooden base, allowing them to discard the pole and glide off with carefree abandon. The technique of pushing off with one foot and gliding with the other is called 'the Dutch roll'.

When Dutch settlers arrived in America, they introduced skating to the masses. The earliest skating centre was in Philadelphia, and the first American ice-skating club was founded there on 21 December 1849. The pastime received worldwide acclaim in the 1850s and 1860s. However, the first steel clamp-on skate was actually invented by Philadelphian-born EV Busnell in 1848. Just over a decade later, American skater Jackson Haines (1840-1875) developed a two-plate, all-metal blade. This was then directly attached to his boots. A 'toe pick' or 'rake' was added by Haines in the 1870s, thus enabling him to

ORIGINAL FACTS

- The oldest pair of skates dates back to 3,000BC. They were found at the bottom of a lake in Switzerland.
- At the height of its popularity across America and Europe in the 1850s and 1860s, 50,000 people per day would skate in Central Park, New York.
- The world's largest skating rink is the Rideau Canal, Ottawa, Ontario in Canada. It is 7.8 kilometres long.

perform the impressive twists, spins and jumps that we associate with figure skating today.

It was during the eighteenth century that ice-skating developed into a worldwide sport. It has been developed into three different fields: figure skating, speed skating and ice dancing, all of which feature in the Olympic Winter Games.

INDOOR TANNING

Natural tanning is caused by the skin's reaction to ultraviolet light from the sun. The first ultraviolet lamp for indoor use was developed by a German medical company called Heraeus in 1904, but it was not intended for tanning purposes. Instead, it was designed to strengthen the bones of patients suffering with rickets. At the time, scientists were aware that sunlight had a beneficial effect on people afflicted with the disease, but the exact mechanism, namely the body's use of UV light to synthesise vitamin D, was not understood until the 1920s. Therefore, the Heraeus lamp was therefore remarkably prescient.

The modern tanning bed was invented in 1975 by a German named Friedrich Wolff. While investigating the effects of artificially produced UV light on the performance of athletes, he noticed that the subjects of his experiments were developing glowing tans. Wolff decided that this welcome side effect had commercial possibilities, so he used his UV lamps

US Patent 4,095,113 13 June 1978

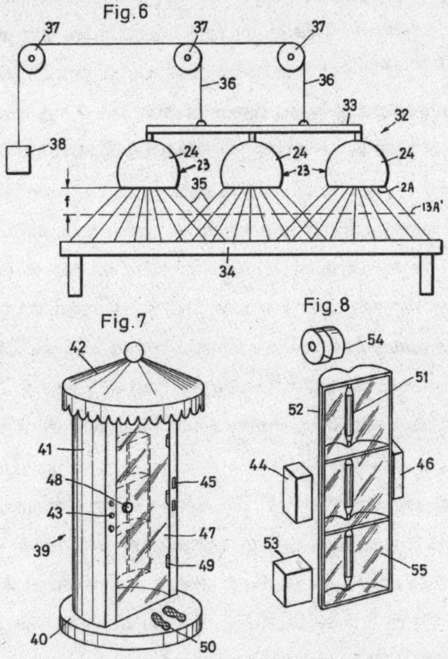

Wolff's original sunbed design, using an array of UV lamps.

to construct a 'sun bed'. The invention sparked off a craze that has yet to subside, despite the dangers that have become increasingly apparent.

INTERNET

The younger generation tends to take the internet for granted; they are probably unable to imagine life without it. But anyone over a certain age will remember its sudden growth in popularity – and will wonder how we managed life without it. The original concept of the internet goes back a lot longer than one might expect: to the 1950s, during the Cold War, when the Soviet Sputnik launch resulted in political tensions that spurred the US to create a 'stable, robust network' capable of linking academic and military computers. The key feature of this network was the way it was structured. Instead of a hierarchal framework of computers, which could be severely damaged or destroyed if the central parts were targeted in an attack, it was composed of millions of nodes, with millions of routes threading through them – rather like a sprawling system of roots. The theory was that if one computer or node in the system was damaged, communications between the rest of the system could still take place.

It is extremely difficult to pinpoint any one person as the internet's genuine 'inventor', because it evolved, and continues to evolve more organically than any other invention. That said, one of the most widely recognised contributors to the internet as we know it was a British man, Sir Tim Berners-Lee, creator of Hypertext Mark up Language (HTML), the code used on the internet, and the World Wide Web, which he called 'an abstract space of information'.

In 1990, the first trials of the internet were undertaken at CERN laboratories (one of Europe's largest research laboratories) in Switzerland. By 1991, the first browsers and web server software was available, and by 1993, the Mosaic browser became the first popularly available browser.On 30 April 1993, CERN's directors announced that 'www' technology would be freely available to everyone, with no fees payable to CERN. This was a milestone in internet history, since it enabled information sharing on a scale never seen before.

Between 1994 and 2000, the internet saw massive growth, unmatched by any other previous field of technology. The first search engines began to appear in the mid-1990s, and it wasn't long until 'Google' dominated the market. Initially, the internet was mainly used for displaying information, online shopping came a bit later. The first large commercial site was Amazon, which began by concentrating solely on books. Today, the internet contains at least four billion pages of information, pumped out by around 50 million hosts to nearly a billion users worldwide. It has created massive new brands (Amazon and eBay to name but two), and the internet shows no signs of slowing down.

J

JET SKI®

Jet Ski is a trademark owned by Kawasaki. The generic term is 'personal watercraft' (or PCW) and there are two basic types: the 'stand-up' and the 'sit-down'. The invention of both is credited to Clayton Jacobsen II, a former banker from Arizona. Jacobsen was a keen motorcyclist who was seized with the idea of translating his favourite pastime to water. In the mid-1960s, he developed a new kind of sea vehicle, which was powered by a jet pump rather than an outboard motor. In 1967, Jacobsen was approached to develop a sit-down version of his machine for the

US Patent 3,369,518 20 February 1968

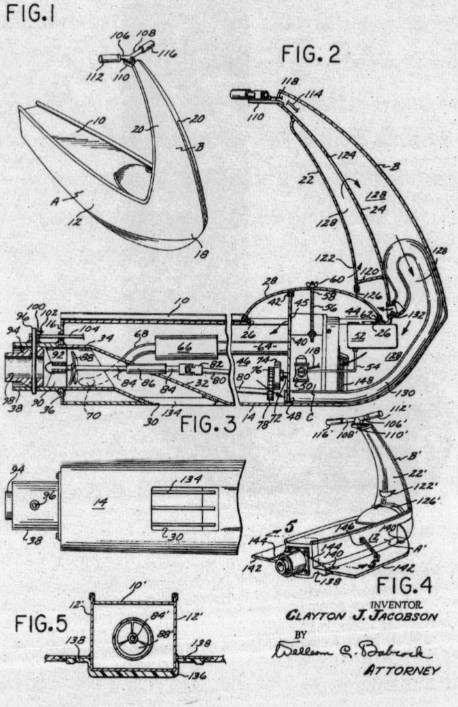

Clayton Jacobsen's original jet-ski design.

Bombardier Corporation, the manufacturer of the Ski-Doo snowmobile. The partnership resulted in the Sea-Doo, which went on sale in 1968. Unfortunately, the product was beset with teething problems and Bombardier dropped the Sea-Doo two years later. At this point, Jacobsen sold his patent rights to Kawasaki, for whom he developed a stand-up PCW called the Jet Ski. In view of its success, Bombardier launched a new version of the Sea-Doo in 1988.

JIGSAW

The origin of the jigsaw is most commonly attributed to an English engraver and map-maker, John Spilsbury, in 1767. It began when he had the idea of mounting one of his maps on hardboard and cutting around the edges of the countries with a fine bladed saw. These early map jigsaws were then used as geography education tools.

Initially jigsaws had the rather scientific name of 'dissections'. The term 'jigsaw' only came about in 1880, when the treadle saw was introduced to cut them. Technically a fretsaw, not a jigsaw was used, but the name somehow stuck. They remained primary educational tools until about 1820, but the introduction of plywood towards the end of the century allowed illustrations to be glued or painted on the front of the wood. Pencil tracings of where it needed to be cut were made on the back.

It wasn't only children who enjoyed puzzling over the jigsaw pieces. The 1900s saw adults become addicted to joining pieces of the puzzles together, but unlike the children's puzzles, these were more challenging and had no guide printed on the box.

Wooden puzzles were a lot more expensive, as they had to be cut one piece at a time. An average 500-piece puzzle cost US$5 in 1908, which made it an expensive toy. The average weekly wage was around US$50.

In 1929, during the Great Depression, puzzles were a piece of escapism. Thanks to the high unemployment rates, home entertainment was a popular option. Many architects and carpenters began making and selling puzzles. Eventually prices were driven down, making it more affordable.

US Patent 1,590,488 29 June 1926

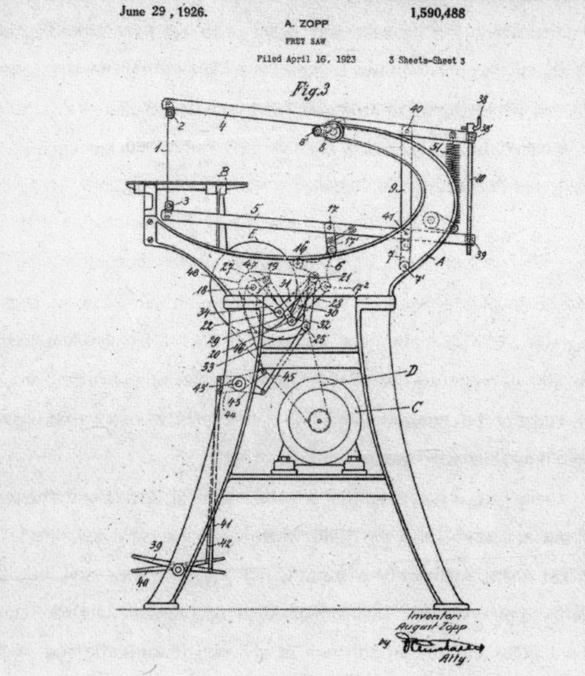

The puzzles were initially called 'dissections'. The term 'jigsaw' only came about when people began using treadle saws (saws powered by a foot-operated lever) to manufacture them. Technically these were fretsaws, but 'jigsaw' was the term which stuck.

The 1920s and 1930s were the golden age of jigsaw puzzles. Mainstream companies such as Chad Valley and Victory in Great Britain and Einson-Freeman and Viking in the US produced a wide range of puzzles. In 1932, a weekly jigsaw puzzle, 'Jig of the Week' was established. It retailed for 25 cents and appeared every Wednesday on newstands. This became a new craze for jigsaw fans who would compete with their friends to finish it first, considering it a real accomplishment.

The era also saw the emergence of what is popularly known as the 'Rolls Royce of jigsaw puzzles'. These were the Par puzzles invented by Frank Ware and John Henriques in 1932. They developed individualised jigsaws, often cutting the customer's name or initials into them, and tried to market their jigsaws to movies stars and even royalty.

After World War II, sales of wooden jigsaws declined, but they did not disappear altogether. Springbok Puzzles produced jigsaws from high-quality art. In 1965, hundreds of thousands of Americans struggled to assemble Jackson Pollock's 'Convergence', which Springbok called 'the world's most difficult jigsaw puzzle'.

A few years later, Steve Richardson and Dave Tibbetts decided to get in on the act. They founded Stave Puzzles and soon succeeded Par as the market leader in wooden puzzles. They experimented with pop-up figures, which eventually led to 3-D puzzles.

It wasn't until the end of the twentieth century that cardboard puzzles could be die-cut. This process allowed thin strips of metal with sharpened edges to be twisted into intricate patterns and fastened to a plate. The plate was then placed in a press and then pushed down to make the cut, a bit like a giant pastry cutter. Jigsaws could therefore be produced far more quickly and efficiently.

Modern jigsaws are commonly made of cardboard, they are quicker to produce and easier to print upon. The most recent development is the Jigtopia jigsaw puzzle, which can be completed on a computer.

K

KALEIDOSCOPE

The Scottish physicist Sir David Brewster (1781-1868) invented the kaleidoscope in 1816. He was a former child prodigy who had built a telescope at the age of 10 then entered the University of Edinburgh at 12. Brewster's main scientific interest was the properties of light. While experimenting with mirrors mounted in metal tubes, he discovered that beautiful symmetrical patterns could be generated by the introduction of fragments of coloured glass. He also developed a mechanism which changed the patterns by shifting the pieces of glass. Brewster's instrument housed two mirrors, which were mounted at an angle of 60 degrees to one another that produced five images. He decided to call it a kaleidoscope, a name derived from three Greek words: *kalos* (beautiful), *eidos* (form) and *scopos* (viewer or watcher). Although Brewster attempted to patent the contraption, there was a problem with his application. Before he could reap the benefits of the insatiable demand generated by his invention, other kaleidoscope manufacturers stepped into the breach.

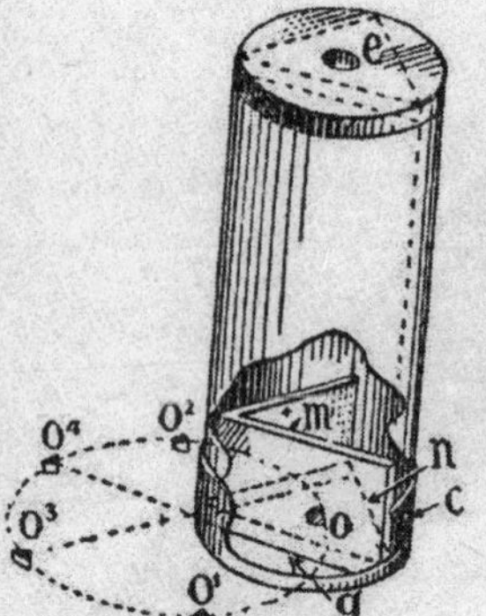

Brewster's kaleidoscopes used fragments of coloured glass and two angled mirrors to produce their symmetrical patterns.

KETCHUP

Today, ketchup is considered as American as apple pie, but it began life as a Chinese and South East Asian dipping condiment. In the Amoy dialect of China, *ke-tsiap* means 'the brine of pickled fish or shellfish', and that is exactly what the sauce originally was. The inhabitants of the region also made sauces from the brine used to pickle vegetables. The tart taste was so popular with English and Dutch traders that they brought *ke-tsiap* back to Europe, where the name was anglicised to ketchup. It was quite a while before the sauce was made with tomatoes. Initially, the most common versions were made from mushrooms or walnuts. An early reference to a tomato-based ketchup occurs in Richard Briggs's *The New Art of Cookery*, which was published in America in 1792.

A 1962 ad for Heinz Tomato Ketchup.

KILT

Seen as a symbol of Scottish heritage, the kilt as we know it, is actually a relatively recent phenomenon. The word kilt literally means 'to tuck one's clothes up', and the original garment was simply a long piece of thick woollen cloth sewn together. Known as the Great Kilt, or the belted plaid, it was gathered into pleats and secured by a wide belt. The upper part could be worn as a cloak draped over the left shoulder. It was worn over a leine (shirt) and discarded when going into battle. The date of its invention remains in debate, but it was certainly around at the beginning of the seventeenth century, because from 1624 it was compulsory to wear a kilt as part of the military uniform.

The short version of the kilt that we know today is attributed to Thomas Rawlinson of Lancashire around the 1720s. The story goes that his workers wore the kilt in his factory in Scotland, but found them too cumbersome to work in. Therefore, Rawlinson proposed cutting them down, essentially removing the top part that would normally be draped over the shoulder. This evolved into the modern kilt, when pleats were sown into it for ease of wear.

KITE

The significance of the kite cannot be underestimated. It first inspired man to fly, and can even be seen as the precursor to the aeroplane. The history of the kite, however, is shrouded in legend, but is presumed to have begun in China, well over 2,000 years ago. Exact stories vary widely, but it is likely that the Chinese drew inspiration from watching the effects of wind on leaves, bamboo hats or sails. They had invented wind sails long before the kite, and it is possible they drew inspiration from these. Interestingly enough, kite flying was not originally a purely recreational pursuit. It is thought early kites were used for military purposes in China, to send messages, measure distances and allegedly carry men over enemy areas. Their recreational use is thought to have derived from the Tang Dynasty when it became a popular pastime.

Kites were introduced to Europe by explorers returning from Asia. In eighteenth-century Europe, the scientific value of kites soon began to be realised. In 1833, a British meteorologist, ED Archibald, used kites to lift anemometers to measure wind speed at various altitudes. This gave a huge amount of information about the atmosphere and dramatically improved weather forecasting at the time. In 1752, the American scientist Benjamin Franklin (1706-1790) flew a kite into an electrical storm and observed sparks coming from a key he had suspended from the line. Therefore, he was able to demonstrate that lightning was similar to static electricity.

Kites were also fundamental in early experimentations with flying machines. In the 1890s, Lawrence Hargrave (1850-1915) experimented near Sydney with a number of kite designs. Through trial and error, he developed the cambered aerofoil, a feature that created much more lift than a flat surface. This formed the basic shape of aircraft wings today. However, it is the Wright Brothers, Wilbur (1867-1912) and Orville (1871-

1948), who are most famous for being the first people to achieve powered and controlled flight, with their design in 1903, which was the culmination of several years experimenting with kites.

KLEENEX®

During World War I, manufacturing company Kimberly-Clark developed a material called 'Cellucotton' as both a surgical dressing and filter for gas masks. At the end of the war the firm was left with a huge surplus of the now apparently redundant cotton substitute. The company's executives scratched their heads to find a commercial use for it. In the end, they decided to use 'Cellucotton' to manufacture make-up-removing facial tissues for women. The Kleenex brand was launched in 1924. By 1926, Kimberly-Clark was receiving letters from disgruntled women complaining that their husbands were using the tissues to blow their noses. Intrigued, the company placed an advertisement in the local paper in Peoria, Illinois, in 1930 to test the market. Readers were invited to choose between two coupons redeemable for a box of Kleenex. The first endorsed the product as a cold cream remover, while the second showed a preference for using it as a handkerchief. Of the respondents, 61% picked the nose-blowing option and Kimberly-Clark altered its marketing strategy as a result.

'Cellucotton' was used for bandages and gas mask filters during World War I. It found its way into people's homes after the war as Kleenex tissue.

KOOL-AID™

Kool-Aid was the brainchild of a mail order salesman from Hastings, Nebraska. Edwin Perkins (1889-1961) spent much of his youth knocking up potions and perfumes on his mother's kitchen table. At the age of 25, he founded Perkins Products Co to market his inventions via magazine adverts. One of the most popular products was a drink concentrate called Fruit Smack. It was offered in six flavours: cherry, strawberry, grape, orange, lemon-lime and raspberry. The problem was that the four-ounce bottles were expensive to post. Perkins thought how nice it would be if all the liquid could be removed ahead of posting, leaving the customer with a lightweight powder that could be easily reconstituted into a refreshing drink. In 1927 he developed a process that did exactly that. He christened the product Kool-Ade (the spelling was changed later) and sold it for a dime a packet. By 1950, the company was making almost a million packets per day.

L

LAPTOP

The first truly portable laptop is thought to have been invented by Adam Osborne who founded Osborne Computer. In 1981 the Osborne 1 was produced; a portable computer that weighed 24 pounds and cost $1,795. It came with a five-inch screen, a modem, software programs and a battery pack. After its initial success, the company soon found it harder and harder to compete. Similar models with superior technology were starting to enter the market.

In 1984, IBM introduced its Personal Portable Computer. At that time, the only manufacturer capable of challenging IBM's supremacy was Apple, which launched its famous '1984' ad for the Mackintosh in a bid to portray its rival as stale and outmoded. Featuring a dystopian scenario of oppressed workers hectored from a giant video screen by a creepy 'Big Brother' figure, the ad was directed by A-list Hollywood filmmaker Ridley Scott, cost US$1.6 million to make and was booked for a single airing during that year's Super Bowl for $500,000. Nevertheless, for anyone smaller than Apple, anything that was not IBM-compatible was bound to fail – and although Osborne produced more up-to-date models, he was unable to keep up with the bigger players.

By the time Apple's ad hit the airwaves, Osborne had already filed for bankruptcy. Since then, the world of the laptop computer has been

The Osborne 1—the world's first portable computer.
Gym membership should have come as standard.

dominated by the likes of Apple and IBM. For the ultimate in minimal extremism, the PDA – Personalised Digital Assistant – was released in 1993 as the first pen-based, hand-held computer. One of the earliest models – the Newton – was issued by, you guessed it, Apple.

LASER BEAM

Albert Einstein predicted lasers in 1916, and it took a further 40 years before his ideas became reality.

'Light amplification by stimulated emission of radiation', of which 'laser' is an acronym, was predicted by Albert Einstein in 1916. It took more than 40 years, however, for the vision to be turned into reality. Charles Townes came close in 1954, when he developed the 'maser' at Columbia University in New York. However, his device produced beams consisting of microwaves rather than visible light. The question of who developed the first true laser is the subject of much controversy. The main candidates are Gordon Gould (1920-2005), Townes' protégé at Columbia, and a young Californian physicist named Ted (Theodore) Maiman (b1927). Gould was the first person to use the term 'laser' and he attempted to patent the phenomenon in 1959, but Maiman is generally considered to have produced the first functional beam. He did this on 16 May 1960, using a silver tipped ruby rod as a light source.

LAWNMOWER

Before the dawn of the lawnmower, lawns were a bushy mass of green splendour. Neat lawns with short grass first appeared in France around the 1700s. They were originally kept clean and tidy thanks to grazing animals, or the occasional snip with hand-held shears. Thankfully, lawnmowers have

Early lawnmowers often required a horse or pony to draw them along and a person to operate them – surely a small price to pay, though, for a mower that 'floats over the uneven ground as a ship rides the waves', as this old Townsend ad promises.

improved and developed since they were first invented in 1830, when the first patent for the design was granted. Realising this was a gardening essential, Edwin Beard Budding, an engineer from Stroud, England, went into business with fellow engineer John Ferrabee, to produce the innovative 'lawnmower'. Budding's discovery was based on a cutting reel, which was used to give cloth a straight, smooth finish after weaving. If he could only think of a way to fix the cylinder onto a frame and make the blades rotate close enough to reach the lawn's surface, he'd be onto a winner... so he did.

Budding and Ferrabee had enough business acumen between them to allow other companies to build mowers under their licence, the most successful of these being Ransomes of Ipswich, England, which began making mowers as early as 1832. The company has made mowers ever since, and is now the world's largest manufacturer of lawn care essentials.

The early lawnmowers were often drawn by horses wearing large leather boots to prevent damage to the grass. Alexander Shanks, from Scotland, was the first to build such a machine. In 1841 he built a 27-inch pony-

DID YOU KNOW?

- The term 'lawnmower' was first recorded in print in 1875, and the word 'mow' derives from the tenth-century word 'mawan', which has the same German etymology as 'meadow'.
- Lawnmower racing was first established at the Cricketers Arms pub in Wisborough Green, West Sussex, England, in 1973 and has been thriving ever since. There are three categories of competition: run behind the mower, tow the mower, or sit on the mower.

drawn lawnmower, and by 1842 he had 'supersized' it to a 42-inch horse-drawn mower.

The first motor-powered lawnmower was built by James Sumner of Leyland, Lancashire, England, at the end of the nineteenth century. It was powered by steam and weighed an astonishing two tonnes.

LEGO™

The versatility of Lego has endured the test of time, and remains a popular toy despite the popularity of computer and video games. It all began in a small village called Billund in Denmark, where Ole Kirk Christiansen (1891-1958) had a factory that produced stepladders, ironing boards and wooden toys. In 1934 Ole adopted the name 'Lego' for his products, formed from the Danish words *LEg GOdt* or 'play well'. Ironically, these words, when translated from Latin, also had an apt meaning: 'I study' or 'I put together' according to the LEGO Group.

The main transformation in the evolution of Lego was in 1947 when Ole's factory was the first to buy a plastic-injection moulding machine for making toys. In 1949, the company started to produce Automatic Binding Bricks – the forerunner to the Lego brick. They were similar to today's toy, but without the tubes found on the inside of the brick.

It was not until the heritage of Lego was passed down to Ole's son, Godtfred Kirk Christiansen, in the 1950s, when he was made junior vice-president, that the company received more orders for plastic toys than for wooden. In 1953 Godtfred perfected the Lego in strength and colour and established the name 'Lego mursten' ('Lego bricks'). Then, in 1958, he invented the interlocking plastic blocks, known as the stud-and-tube coupling system. His newly modelled Lego bricks received high praise and were considered safe and creative by adults.

The 1960s saw Lego's market expand to countries such as Norway, Germany, the UK and Switzerland. A few years later, Lego was introduced to the USA and the first sets came out with loose bricks. In 1964, instruction manuals were included with the Lego sets to help children create buildings, planes and other constructions.

On 7 June 1968, Legoland Billund opened its gates. Nowadays, it attracts millions of visitors per year, with an outstanding 1.63 million visitors in 2003. Legoland Billund is now much bigger and attracts many more people owing to Ole Kirk's legacy. It has been called a 'child's paradise', where the imagination can run wild. The Windsor Lego Park opened in the UK in 1996, using an amazing 25 million Lego bricks in its construction.

The popularity of Lego improved a great deal in 1978 when Legoland mini figures were introduced. These tiny 'people' had moveable arms and legs and printed faces. At the same time, Lego sets were expanding thanks to the addition of a variety of baseplates and road plates that allowed an entire town to be built. Today, there is even a line of superhero Lego figures called Bionicles. Loved by successive generations, Lego looks primed to live up to its innovative past and continues to move with the times.

LIGHT SWITCH

The question of who invented the light switch does not have a straightforward answer. Electric lights were made by Humphry Davy (1778-1829), Heinrich Göbel and Alexander Edmond Becquerel (1820-1891) of France between 1801 and 1880, when Thomas Edison (1847-1931) clinched the patent for his bulb design aimed at the general public. As such, it follows that any of the four could have created a mechanism with which to spark up

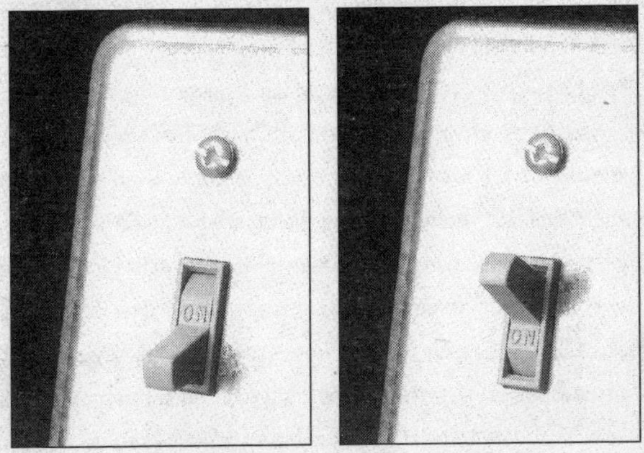

UK, Australia, New Zealand

United States

Anything to be different

Although a very simple invention, the light switch still manages to vary from country to country.

their creations. All commercial switches did was to miniaturise the workings, which were pretty simple anyway because they only had to connect a couple of contacts.

Light switches have undergone very few transformations or variations over the years; even their appearance has been tampered with surprisingly little. One reason for this may be that, as they're generally durable, they are replaced only when a house is rewired, which happens pretty infrequently. Before 1970, some light switches contained a glass vial of mercury, which was tipped by the switching action: the mercury rolled down the angled vial, bridged a pair of contacts to complete a circuit and allowed the light to turn on. These kinds of switches sometimes glowed in the dark to make them easier to locate. By comparison, the modern toggle or rocker switch uses an arrangement of springs and levers to get the contacts talking to each other.

Another light switch innovation was Granville Woods's dimmer switch, which was invented in the 1890s to provide better control in theatre lighting. It was eventually adopted by the domestic market. The dimmer switch was followed by the sound sensitive IntelaVoice switch, invented by Rick Matulich of VOS Systems, which requires only a handclap or single word to be activated. Oddly enough, this switch has yet to take off in the mass market, even though it allows people to swan around in the privacy of their own homes, bossing their household electrics about.

The latest development in the realm of light switches looks to be a massive advancement on anything that has preceded it. Using the protein FP595 – present in the fronds of a Mediterranean coral – scientists are working to create the molecular light switch, which will enable optical data to be stored in compact protein crystals. This could spell the end for that clunky old microfiche in your public library.

Even when an invention has only two positions, this doesn't always guarantee consistency. As all light-switch fanatics know, in the UK, Australia and New Zealand, the 'down' position means 'on', whereas in the US, 'on' is in the 'up' position. The same fanatics will probably enjoy the Thomas Edison light-switch cover from the Philosophy Gift Shop – operated by cafepress.com – featuring a black and white photo of the great man beneath a dust-proof laminate.

LINOLEUM

In 1860, the English rubber manufacturer Frederick Walton patented a new kind of floor covering. He was inspired by the skin that had formed on the surface of a can of paint in his basement. Walton realised that the thick, rubbery substance was oxidised linseed oil. He mixed the paint with ground cork and spread it onto a piece of canvas. When it dried, he was delighted with the results. He realised that the product would make an excellent, durable and cheap alternative to Kamptulicon, the vogue flooring material of the day. Walton decided to call his invention 'linoleum', a name derived from the Latin words for 'flax', the plant that yields linseed oil, and the oil itself.

Walton patented linoleum on 19 December 1863 and it soon became a standard fixture in Victorian homes. Linoleum fell out of favour with the arrival of synthetic alternatives during the 1960s, but it has recently enjoyed a renaissance. It is an eco-friendly, naturally fire retardant material and is available in a great deal more vivid colours than it once was.

LIPOSUCTION

Since its invention in 1974, liposuction has become the United States' most commonly performed cosmetic surgery procedure. The technique was pioneered in Rome by an Italian gynaecologist named Dr Giorgio Fischer. He used an oscillating blade to cut away at fatty tissue, which was then suctioned out by a vacuum tube along with a sometimes copious quantity of blood. This so-called 'dry' operation, in which no fluids were injected into the targeted area of the body, frequently led to complications such as numbness and post-surgical pain. These side effects were significantly ameliorated when the Californian dermatologist Jeffrey Klein introduced 'wet' or tumescent liposuction in 1985. This entailed swelling the tissue ahead of surgery by injecting it with saline solution containing epinephrine (a blood vessel constrictor) and lidocaine (a powerful local anaesthetic). This resulted in much less nerve damage and loss of blood.

LIPSTICK

In 1999, a team led by the archaeologist Christopher Henshilwood of the State University of New York found a curious object in Blombos Cave on the Indian Ocean coast of South Africa. It was a three-inch piece of red ochre, an iron-based mineral traditionally used by local tribes to make body paint. The item had been deliberately cross hatched with an abstract design and its shape was strikingly reminiscent of a stick of lipstick. The following year, the team discovered another one. When the objects were dated using a technique called thermoluminescence, they were found to be approximately 70,000 years old. The evidence is not conclusive, but several anthropologists believe that the Blombos finds are the world's oldest examples of lipstick.

M

M&M's®

On a trip to Spain during the Spanish Civil War (1936-1939), the confectionery magnate Forrest Mars Snr (1904-1999) noticed some troops eating small chocolates coated in hard sugar. They were delicious, texturally interesting and their sweet shell prevented them from melting in the fierce Iberian sun. Mars decided to produce similar candies for the American market. The first tubes of M&M's were sold in 1941. The first 'M' in the chocolates' name is easy enough to explain – it stands for Mars – but the second is more surprising. It is rumoured to represent the surname of Bruce Murrie, the son of the president of Mars's chief competitor Hershey. The explanation for this unlikely tribute is that Murrie supposedly provided 20% of the finance for M&Ms. Mars was keen to get him involved because Hershey had a contract to make sweets for American troops involved in World War II. At a time when chocolate was rationed, this joint venture secured Mars access to its most important ingredient.

MACINTOSH RAINCOAT

An essential wet weather item, the common rain mac is named after the Scottish chemist Charles Macintosh (1766-1843). In the process of trying to work out what could be done with waste products from gasworks, Macintosh discovered that coal-tar naptha dissolved India rubber. The first practical waterproof material was a rudimentary combination of two pieces of material sandwiched together with rubber and softened with naphtha. Originally, the intention was to make tarpaulins, but tailors started to use the new material for making raincoats. However, this did not prove to be particularly successful, as sewing the fabric let the rain in through the seams. Therefore, Macintosh developed waterproof seams. The original raincoats were hard to miss – they were the bright yellow kind that American weather forecasters used to wear. Early macs were not only quite eye-catching, but their aroma was pretty strong, too. When it rained, the odour they gave off could be detected halfway across the street. It was only as technology improved that odour-free variations emerged. Today, the rain mac has evolved into the raincoat and the trench coat and, thanks to labels like Burberry, have even become fashion items.

MARBLES

Marbles have had a variety of uses throughout the ages; they were first enjoyed in the ancient world. In Ancient Greece and Rome, youths played with clay balls for entertainment, and marbles have even been discovered in the tomb of the young Egyptian pharaoh, Tutankhamun. During the Middle Ages, marbles continued to arouse fascination, although children who played with them were seen as delinquents. By Elizabethan times, marbles were still being made of clay and used for recreation. For long-buried linguistic reasons – that had nothing to do with a distaste for Socialists – marbles were often referred to as 'commies.'

It was in Germany that marble production really took off – hence their name, which means 'from the rock' in German. By 1775, water-powered stone mills were producing polished pieces drawn from 26 agate mines in the vicinity of Ida Oberstein. Around 1802, when agate supplies were thought to be dwindling, the industry adapted with remarkable speed,

Although mainly collectibles now, marbles were for many years an extremely popular children's toy.

forging import links with India, and later with Uruguay. Production increased in the 1860s, and it peaked in the 1880s, after which Germany continued to lead the field until the dawn of World War I. From this point, the majority of stocks were exported to the US, and the industry standard shifted from stone to glass.

Germany was largely responsible for this, too. In 1846, a native glassblower invented *marbelschere* (marble scissors), which were used in conjunction with a cup that used heat to shape glass. However, it was America that took the lion's share of marble production, thanks to Martin Christensen (1849-1915). Using the barn behind his home in Ohio, Christensen manufactured masses of marbles with the aid of the first-ever marble machine. By 1910, his workers were making up to 10,000 marbles a day. But Christensen's gravy train came to an abrupt halt when the natural gas that he used to fire his furnace was rationed at the onset of World War I.

Ohio, though, had not seen the end of marble production. Sam Dyke, owner of the company Akro Agate, and a marble producer since 1880, established himself as the chief marble manufacturer for the next few decades. Despite Christensen's problems, it was his innovative marble machine that helped Dyke's enterprise prosper. Then, in 1950, a new kind of marble emerged from Japan. It was made by injecting coloured glass into the middle of a clear orb, and enigmatically dubbed the 'cat'seye'. The modern marble had arrived.

Marbles are now hugely coveted collectibles – the older they are, the more they are highly prized. They have also rolled into the language of popular culture, when astronaut and third man on the moon, Charles P 'Pete' Conrad quipped that 'the Earth resembled a beautiful, blue marble suspended against a black velvet blanket'. The terminology of marble playing has also entered everyday usage: 'playing for keeps', 'losing your marbles' and 'knuckling down' are all direct quotations from the game. While technological advances in toymaking have made marbles a less popular choice than they once were, as collectables, they will be filling people's pockets for a long time to come.

MASCARA

An essential tool in any woman's make-up bag, mascara derives from the Italian term *maschera* – meaning 'mask'.

There are conflicting reports over the creation of the first modern mascara in New York in 1913. Some say either that it was dreamt up by chemist TL Williams for his sister, Mabel – to stop her boyfriend, Chet, from straying to another woman – or that it was Mabel's idea all along, and that her cunning brother simply stole it. Whatever the case, the winning blend of Vaseline and coal dust eventually had the desired effect: Mabel and Chet were married the year after mascara was invented.

In 1915, Williams combined his sister's name with the word 'Vaseline' to create the brand name of Maybelline. The brand was soon a mail order hit, and was already all the rage by the time it made its debut at American chemists in 1917.

The Maybelline product relied on modern ingredient Vaseline, but mascara actually had its roots in ancient culture, and was widely used in various forms in Egypt, Greece and Babylon. Galena, lead sulphite, malachite and kohl were just a few of the more exotic ingredients used thousands of years ago, and evidence of such tints survives in a now increasingly popular female name from that period: Keren, which derives from the name of Job's youngest daughter, Kerenhappuch, which means 'horn of antimony' in Hebrew. Antimony, a metallic element, was a prized ancient eyelash paint, so Kerenhappuch effectively translates as 'horn of mascara', and no doubt became a girl's name because of its links with female grooming.

Though made with different ingredients, mascara was widely used as far back as ancient Egypt, Greece and Babylon.

MATCHES

Irish physicist Robert Boyle (1627-1691) developed the first, somewhat rudimentary, friction match in 1680. It consisted of a sulphur-tipped shard of wood and a piece of rough paper coated with phosphorous. When the match was drawn against the paper, a chemical reaction between the phosphorous and the sulphur caused the tip to ignite. As phosphorous was a rare commodity, however, Boyle's invention was regarded as little more than an expensive novelty.

In 1827, John Walker (1781-1859), an English chemist, discovered that if he coated the end of a stick with certain chemicals (antimony sulfide, potassium chlorate, gum and starch), he could start a fire by striking the stick anywhere. He didn't patent his 'friction matches', which he called 'Congreves' and made little money out of them. The first commercially viable matches were the 'Lucifers' manufactured by a Londoner called Samuel Jones (b1801) in the latter part of the 1820s. They were popular with smokers, but they produced an atrocious smell on ignition. This prompted a French chemist named Charles Sauria (1812-1895) to introduce a phosphorous-based alternative in 1830. Sauria's matches were less malodorous, but they had two serious drawbacks. First, they were too easy to light. The slightest friction caused them to flare up; they caused innumerable fires and injuries. Second, their main ingredient was highly poisonous. Workers in match factories were routinely afflicted with 'phossy jaw', a horrendous necrosis of the bones. Many babies died, or were deformed, as a result of ingesting phosphorous through sucking match heads.

Don't play with fire:
From Sauria's overly flammable friction matches through to today's safety matches, there has always been the potential for disaster – as this old warning ad tries to convey.

Nevertheless, the public appetite for smoking was such that the risks were tolerated, pending the invention of a safer alternative. This breakthrough occurred in 1855, courtesy of the German chemist Anton von Schrotter (1802-1875). He resurrected Boyle's idea of dividing the combustible components between the match heads and the striking surface. The result was the safety match, but the term is only relative: it remains a very bad idea to eat the tips.

MAYONNAISE

There are several theories about the origin of mayonnaise. Some say that the emulsion of egg yolk, oil and vinegar takes its name from the French verb *magner*, meaning 'to stir' or 'make by hand'. Others believe that the sauce was first made in the French city of Bayonne and that the 'b' somehow got replaced by an 'm'. Still others hold that mayonnaise was named after the Duc de Mayenne, who famously insisted on finishing his meal of chicken and cold sauce prior to the Battle of Arques in 1589. (He was subsequently defeated by Henri IV.) The most popular explanation, however, is that the sauce was first whipped up in 1756 to celebrate the French capture of Mahon, the capital of the Mediterranean island of Minorca. Forces led by the Duc de Richelieu had wrested control of the town from the British. The victory banquet was supposed to include a sauce made of cream and eggs, but Richelieu's chef was unable to find any cream. He decided to substitute olive oil, and Mayonnaise (or Mahonnaise) was born.

MECCANO™

Meccano, the internationally popular mechanical construction system, was the brainchild of Frank Hornby (1863-1936), a young clerk from Liverpool. While building simple mechanical models for his children, Hornby became frustrated that each new model required a new set of parts; there was no possibility of reusing the parts to create different designs. It struck him that interchangeable parts, which could be assembled differently to create a variety of models, would allow children much more room for experimentation and would be far more cost-effective to produce. Hornby patented his system, 'Mechanics Made Easy', in 1901 and the Meccano trademark was registered in 1907. It is thought the word 'meccano' comes from the phrase 'make and know', reflecting Horby's keen interest in the education of his children.

Whatever, the craze soon caught on. The first Meccano magazine appeared in 1916, and the worldwide Meccano Guild and Correspondence Club was founded in 1919. This was the start of the Hornby Toy empire. By the time Hornby died in 1936, he had overseen the creation of train sets; electrical, chemistry, aeroplane, conductor and radio sets; toy speedboats and Dinky toys. With his Meccano idea, he had taken a simple premise and turned it into a versatile system which could replicate virtually any mechanical device. Frank Hornby leaves behind him an international community of Meccano enthusiasts as testimony to his invention.

Thomas Edison invented the modern megaphone, although his original device also featured two funnels designed to aid hearing as well.

MEGAPHONE

According to legend, Alexander the Great used to address his armies through an enormous sound-amplifying funnel. This basic principle has been in use since people first cupped their hands around their mouths to increase the volume of urgent messages. Strangely, though, the modern megaphone owes its existence to a device primarily designed to aid listening. It was built by master of communication Thomas Edison (1847-1931). His invention comprised two tapering horns about six feet long that fitted with ear tubes at their narrow ends. The horns were mounted on a tripod, together with a large speaking trumpet and allowed conversation over a distance of two miles. Edison exhibited his megaphone at the Paris Exhibition in 1878. The ear tubes eventually became redundant when it became apparent that the typical user was only interested in one-way communication, usually for purposes of crowd control or propaganda. The other major development was the introduction of an electronic amplifier, whereupon the megaphone became the bullhorn.

METAL DETECTOR

Scottish inventor Alexander Graham Bell (1847-1922), best known as the father of the telephone, also built the world's first metal detector. His motivation was to save the life of President James Garfield, who had been shot by Charles Guiteau on 2 July 1881 while waiting to board a train. After failing to locate the bullet, Garfield's physicians called in Bell. He rigged up a metal-detecting device – which he called an 'induction balance' – and headed for the White House. Bell scanned the dying President's body several times, but was unable to find the bullet. The reading he did manage to obtain was probably triggered by a spring in the mattress of Garfield's bed.

The hand-held metal detector was patented in 1937 by a German immigrant to the US called Gerhard Fisher. He had stumbled on the necessary technology several years earlier while developing an aircraft direction-finding system. It worked well, except in areas where there were ore-bearing rocks. Fisher realised that the design could be modified to build metal detectors. His invention proved extremely useful for the detection of landmines during World War II. At the end of the conflict, the machines found their way into army surplus shops and were snapped up by amateur treasure hunters.

US Patent 2,066,561 5 January 1937

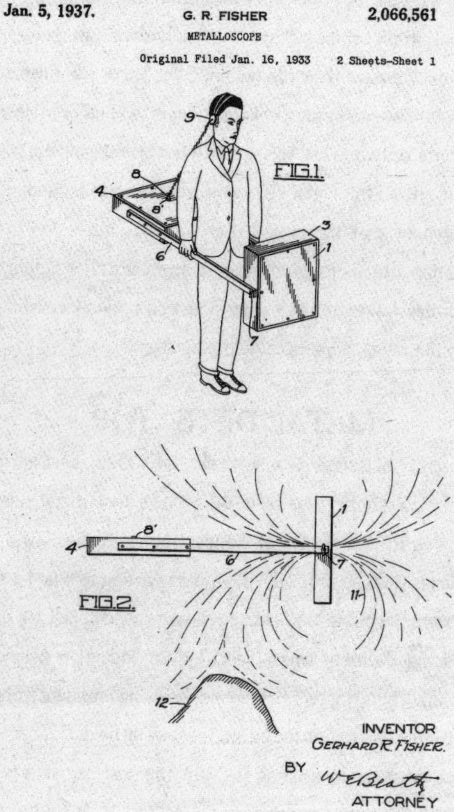

Gerhard Fisher's 1937 'Metalloscope', the first handheld metal detector to be patented.

MICROSCOPE

The first crude microscope was arguably built by Dutch spectacle-makers Zacharias and Hans Janssen during the 1590s. If Zacharias is to be believed – he only announced the supposed invention several years later – it consisted of two lenses mounted at either end of a metal tube. Such devices were certainly in circulation by the early seventeenth century and were known as 'flea glasses'. Whether or not they qualify as microscopes, however, is debatable. They were only capable of magnifying objects 10 times.

The individual generally acclaimed as the father of microscopy is another Dutchman, Anton van Leeuwenhoek (1632-1723). He became interested in magnifying lenses while working as an apprentice in a draper's shop, where they were used to count threads in fabric samples. Leeuwenhoek taught himself to grind his own lenses, and produced examples much more highly curved and polished than any that had been made previously. He used them to build microscopes with magnification powers of up to 270. He then started to look at various substances through the eyepieces. When he examined bodily fluids and samples of pond water, the results were mind-blowing. A universe of micro-organisms was suddenly revealed.

Anton van Leeuwenhoek, generally regarded as the father of microscopy and a pioneer in bacterial research.

In 1673, Leeuwenhoek began to communicate his findings to the most important scientific body of the day, the Royal Society in London. One letter described his microscopic analysis of his own saliva: 'In the said matter there were many very little living animalcules, very prettily a-moving. The biggest sort... had a very strong and swift motion, and shot through the water (or spittle) like a pike does through the water. The second sort... oft-times spun round like a top.' Leeuwenhoek had discovered bacteria. This discovery and his subsequent reports on saliva and blood changed the world.

MICROWAVE OVEN

The microwave oven was accidentally invented in 1946 by an electronics expert at the Raytheon Company in Waltham, Massachusetts. While working on a piece of radar equipment called a magnetron, Percy LeBaron Spencer (1894-1970) noticed something strange going on in his pocket. It transpired that the candy bar he had placed there earlier had started to melt. Spencer guessed that the cause was microwave radiation from the magnetron. Intrigued, he fetched a bag of corn kernels and placed them in front of the device. They started to pop. When he repeated the experiment with an egg, it exploded over one of his colleagues. Raytheon moved

US Patent 2,495,429 — 24 January 1950

Jan. 24, 1950 — P. L. SPENCER — 2,495,429

METHOD OF TREATING FOODSTUFFS

Filed Oct. 8, 1945

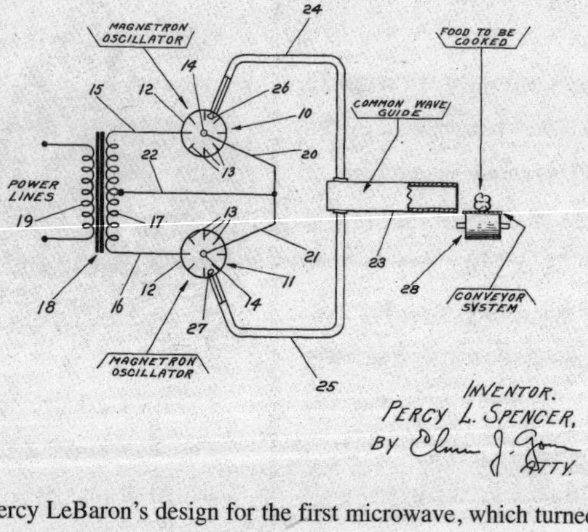

Percy LeBaron's design for the first microwave, which turned out to be a six-foot tall monster costing US$5,000.

rapidly to patent the process and manufactured the first commercial microwave oven the following year. It was a monster, almost six feet tall, weighing 750lbs and retailed for around US$5,000.

MIRRORS

The first mirrors were made from highly polished stone or metal. People in Mexico were looking at their reflections in black obsidian, a form of glass produced naturally by volcanic activity, as early as 4,000BC. Within 1,000 years, the Ancient Egyptians were using bronze mirrors. The first metal-backed glass mirrors appeared in the late twelfth and early thirteenth centuries, and during the Renaissance. Nuremberg and Venice were important centres for the new industry. Venetian craftsmen used tin and mercury to make reflective backings. The method involved covering the glass with the liquid metal before laying it on a tin plate and weighing it down. With skill and luck, there wouldn't be too many blackened blemishes where the mercury failed to adhere. This problem was solved in 1835 when German chemist Justus von Liebig (1803-1873) invented a chemical process for coating glass with silver.

MONOPOLY®

The origins of the property game Monopoly are shrouded in controversy. The story begins in 1904, when a Quaker named Lizzie Magie patented a board game called 'The Landlord's Game'. It had a moral and educational purpose: Magie belonged to a group that campaigned for a federal property tax to discourage land speculation. She wanted to show how the rent system tended to make property owners richer, and their tenants poorer. The game differed from modern Monopoly in that properties were rented rather than bought, but the similarities were striking. The Landlord's Game spread through the Quaker community, with each family adapting the rules and naming properties as they saw fit. In the early 1930s, a woman named Ruth Hoskins brought the game to Atlantic City in New Jersey. Here, it was picked up by Charles B Darrow (1889-1967), who

US Patent 748,626 5 January 1904

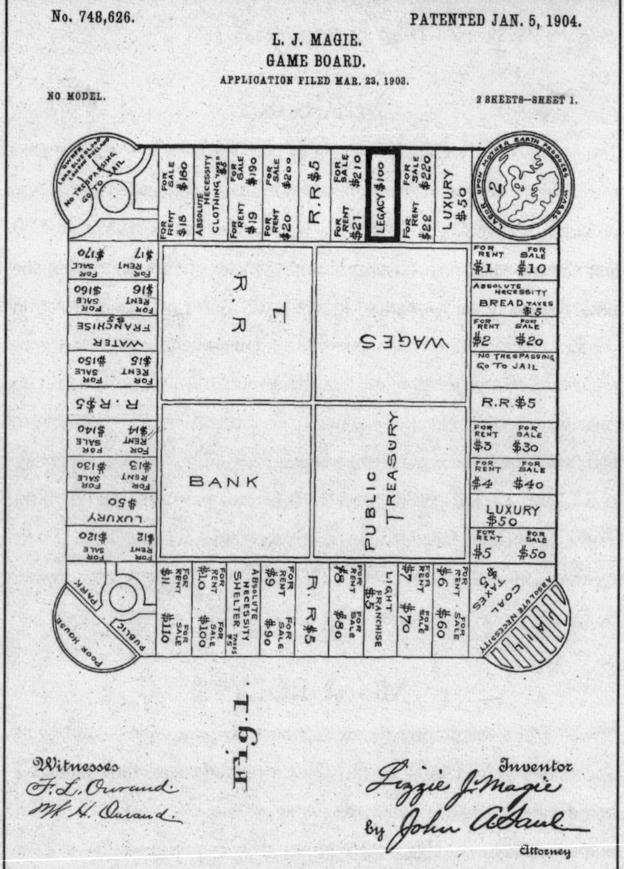

Lizzie Magie's 'Landlord Game' patent, which Parker Brothers bought up for $500 before releasing Charles Darrow's strikingly similar Monopoly game.

busied himself perfecting the game on the family tablecloth. In 1934, Darrow showed his version of Monopoly, which was based on the street names of Atlantic City, to the toy manufacturers Parker Brothers. They rejected it on the grounds that it had 52 design errors. The company relented the following year. Parker Brothers bought up Lizzie Magie's patent for $500 and began to manufacture the first of some 200 million sets sold to date.

It's here where things get complicated. According to Ralph Anspach, a retired economics professor from San Francisco, Darrow was in no position to claim rights to a game that had been in the public domain for 30 years. He also maintains that the name Monopoly was in use long before 1934. Anspach has waged a protracted legal battle with the owners of the brand (first General Mills, now Hasbro) disputing their right to monopolise the name Monopoly. In the early 1980s, the US Court of Appeals ruled in favour of Anspach, a decision upheld by the Supreme Court in 1982. Since then, however, Congress has revised trademark law in Hasbro's favour. Anspach, who manufactures his own provocatively named game called Anti-Monopoly, is now required to pay a license to the company for commercial use of the second part of the title.

MOTEL

The world's first motel opened for business at 2,223 Monterey Street, San Luis Obispo, California on 12 December 1925. It was designed and built by a Los Angeles architect named Arthur Heineman, who planned to build a chain of 18 similar institutions with his business partner James Vail. There were already motor hotels in existence, but it was Vail who thought of contracting the words to come up with the term 'motel'. In view of the pair's ambitions, the San Luis Obispo branch was named 'Milestone Motel', but the Great Depression scuppered their plans, and the putative chain ended up having only one link. As a result, they eventually renamed the hotel 'The Motel Inn'. The Motel Inn consisted of a series of two-room bungalows, each equipped with a kitchen and adjoining private garage. The units were clustered around a central courtyard, which featured a swimming pool and several picnic tables, and the tariff was a modest US$2.50 per night.

N

NAIL VARNISH

Originally made from natural pigments – plus gum arabic, gelatin, egg whites and beeswax – nail varnish was first invented in China around 3,000BC. The Egyptians used a simpler method of dipping fingernails in henna. Traditionally, a woman was only able to wear varnish on entering young adulthood, because of its sexual allure.

The colour of varnish also played a large part in signifying the social classes of China and Egypt. For instance, during the Chou Dynasty (600BC), the royal colours of silver and gold could only be worn by royalty, while the lower-ranking ladies were relegated to wearing pale pink tones. As the colours of royalty changed, so too did nail colour, and black and red later replaced silver and gold. Any lower-class female brazen enough to defy the rules of nail colour faced the death penalty.

These days, the nail varnish industry not only produces a plethora of different colours, but also an impressive array of protective polishes. These include varnishes that strengthen the nails and prevent splitting, as well as those that promote nail growth. There are also protective top and base coats, which help to make the colour last longer and prevent chipping.

It is also worth mentioning that nail varnish was historically a female domain, but since the glam rock scenes of the late 1960s and early 1970s, more and more men began to get in on the act.

In 2003, the first ink nail printer 'NailJet Pro' was released. It allowed custom hi-resolution colour images to be printed onto nails. Other notable nail innovations include jewelling and piercing the finger nails once the polish has dried.

NEOPRENE®

April 1930 was a productive month for the DuPont Company. Its scientific research team, lead by chemist Dr Wallace Hume Carothers (1896-1937), invented not one, but two synthetic materials that are now household names: nylon and neoprene. The latter was the team's first discovery, thanks to the advice of Reverend Julius Nieuwland (1878-1976), a priest and organic

DID YOU KNOW?

- Neoprene was first trademarked by DuPont as 'Duprene'. The choice to change it to 'neoprene' in 1937 was made in order to signify that the material was as an ingredient rather than the product itself.

- Not a drop of neoprene could be found in America during World War II. It was being used for tyres, fan belts, hoses, seals and vehicle gaskets.

- Neoprene is a wonderful insulator. If you ever need a wetsuit, find one made of neoprene. It reduces heat loss by trapping a layer of water between your skin and the wetsuit, while your body heat warms the water.

chemistry professor at the University of Notre Dame, who had taken important steps towards creating a synthetic rubber in 1923. When neoprene began to be used successfully in the production of tyres in 1934, Nieuwland refused to share the royalties; as a priest, he had taken a vow of poverty.

More resistant to oils, water and chemicals than natural rubber, neoprene became a man-made replacement for its natural counterpart. It behaved better as well. Using a process called 'vulcanisation', which combines sulphur with neoprene's long chains of molecules to create links between them, the material could be carefully manipulated to fit the elasticity and durability required of each particular usage. The diversity of products that utilise the 300,000 tons of neoprene produced per year today is remarkable, ranging from car tyres to gloves and Wellington boots. But it wasn't always such a consumer-friendly product. In its initial stages it was plagued with a disagreeable odour. It took a considerable amount of experimenting by Carothers and his team to create the neoprene that is used in everyday products today. The material has remained essentially unchanged since 1950.

NEWSPAPER HOROSCOPES

Newspaper horoscopes were inadvertently started by Queen Elizabeth II's (b1926) younger sister Princess Margaret (1930-2002). In August 1930, the editor of the *Sunday Express* commissioned the astrologer RH Naylor to cast a horoscope for the newly born princess for publication in the paper. Naylor predicted that 'events of tremendous importance to the Royal Family and the nation will come about near her seventh year'. Shortly before Margaret's seventh birthday, the British monarchy was indeed plunged into deep crisis by the abdication of her uncle, King Edward VIII (1894-1972), which also led to Margaret's father, Albert, becoming monarch. Naylor's article was so well received that the editor invited him to write another column for the following week's edition. In it, the astrologer warned that British aircraft were in danger. On the same day the R101 airship crashed in northern France. Naylor became an overnight celebrity and daily horoscopes became a journalistic staple.

NICOTINE GUM

Nicotine gum was the first nicotine replacement therapy product. It was developed by the pharmaceutical company Pharmacia in 1971 and was first sold in Switzerland in 1978. The US Food and Drug Administration (FDA) approved the gum as a prescription-only smoking cessation aid in 1984. It was marketed by SmithKline Beecham under the name 'Nicorette'. Twelve years later, perhaps realising that people inclined to develop an addiction to nicotine were unlikely to use the gum as a first port of call, the FDA permitted it to be sold over the counter.

NICOTINE PATCH

In 1979, an assistant professor of psychology at New Mexico Tech named Dr Frank Etscorn III spilled some liquid nicotine on his arm while conducting research into vomiting. He vomited all right, for eight hours, but the incident also gave him an important idea. Why not incorporate nicotine into the kind of patch used to help sufferers from motion sickness? Etscorn's first attempt to patent the nicotine patch was rejected, but he successfully reapplied in 1986 and subsequently sold the rights to the Ciba-Geigy Pharmaceutical Company in Switzerland. A keen astronomer, he used some of the proceeds to help his university build an observatory.

The first nicotine patch to obtain FDA approval for sale in the US was developed by the Duke University researcher Jed Rose during the early 1980s. It was manufactured by Pharmacia and marketed as 'Nicotrol' by Johnson & Johnson and as 'Nicoderm' by SmithKline Beecham. The patch was available by prescription from 1991, and over the counter from 1996. Despite the success of his invention, Rose is under no illusion that there is more to giving up smoking than just simply finding an alternative source of nicotine. 'Most people think of craving and addiction as a craving for the drug effect *per se*,' he has said, 'but it's really a craving for the whole behaviour, especially with something like cigarette smoking that presents a rich context of sensory cues and habit components.'

NUTRASWEET®

NutraSweet, or aspartame to give it its scientific name, was invented in 1965 by Dr James Schlatter, a chemist at GD Searle & Company. One day while working on an amino acid-based treatment for stomach ulcers, Schlatter licked his fingers to pick up a piece of paper. He was unaware that he had got some of the compound on his hands. His finger tasted deliciously sweet. Research revealed that the substance was around 200 times sweeter than sugar. It also had none of the bitter aftertaste associated with other artificial sweeteners. After years of testing, the FDA approved aspartame for use in food in 1981.

NYLON

What is 'strong as steel' and 'as fine as spider's web'? In case you hadn't guessed, it's nylon. These accolades were the core of the DuPont Company's advertising campaign for the synthetic fibre's most popular application: nylon stockings. The DuPont Company unveiled its largest moneymaker of all time at the 1939 New York World's Fair before an audience of 3,000 enthusiastic women's club members.

Nylon's first uses were for fishing line, surgical sutures, toothbrush bristles, and, of course, women's hosiery. The latter first went on sale to the general public in May 1940. In their first year on the market, 64 million pairs were sold. Unfortunately, the onset of World War II meant that nylon was diverted to military uses, which included parachutes, airplane tyre cords, and glider towropes, but as the conflict came to a close, so did the shortage of nylons. Shoppers flooded into stores demanding the new product. One location in San Francisco was even forced to halt sales because of a mob of 10,000 crazed stocking-seekers. Newspapers ran stories with headlines such as: 'Women Risk Life and Limb in Bitter Battle over Nylons.' Elitist 'silk stockings' were now a thing of the past, but even the price of nylons was still out of reach for most people; thanks to the tremendous demand, DuPont could afford to set his price 10% higher than that of silk.

It may be difficult for us to imagine rioting over something like stockings, but at the time the new product was seen as a miracle of modern science. We owe it all to Dr Wallace Hume Carothers: a scientist so devoted to his calling he awoke before dawn to face the 'terrific grind', likening himself to 'the child labourers in the spinning factories and the coal mines'. His research into large molecules with repeating chemical structures, called polymers, at DuPont's Experimental Station in the early 1930s, led to the discovery of the world's first synthetic fibre.

The objective at DuPont at the time was to promote 'fundamental' research; in other words, to create new scientific facts instead of applying existing scientific knowledge to current problems. DuPont recruited Carothers, who was a professor at Harvard, to work on research into the emerging field of polymer science. His work involved combining the chemicals amine, hexamethylene diamine and adipic acid to create a new fibre through what is known as a condensation reaction – the joining of molecules with water as a by-product. He then modified the process by distilling the water, which had been weakening the strength of the fibres,

A Pretty Polly ad shows nylons have gone from strength to strength.

and there you have it: a synthetic substitute for silk, patented as 'nylon'. The process by which the name was chosen was as follows: norun (a bit too presumptuous), nuron (too scientific), nulon (they felt 'new nulon' was considered a bit redundant), nilon (no fewer than three possible pronunciations), and finally nylon. The company chose well; the term 'nylon' is now a household name.

O

ORGAN TRANSPLANT

Many people know that Dr Christian Barnard (1922-2001) performed the first successful human heart transplant in 1967. Less well known is the date and nature of the first organ transplant of any kind. It happened in Olomouc in Moravia (now part of the Czech Republic) on 7 December 1905 and involved the transplantation of the corneas of an 11-year-old boy named Karl Brauer (b1894) into the eyes of Alois Glogar, a labourer who had been blinded the previous year while slaking lime. The operation was carried out by the Austrian surgeon Eduard Zirm (1863-1944). It was partially successful: one of the transplants was rejected but Glogar's other eye worked well enough to allow him to return to work.

The first successfully transplanted internal organ was a kidney donated by Ronald Herrick to his identical twin brother Richard on 23 December 1954. The surgery was performed by Dr Joseph Murray (b1919) at the Peter Brent Brigham Hospital in Boston, Massachusetts. Richard lived for another eight years; at the time of writing Ronald is still alive.

P

PACEMAKER

The cardiac pacemaker was invented by a Canadian named John Hopps (1919-1998) in 1950. There was just one problem with Hopps's device: it was too big to fit into patients' bodies. The first internal pacemaker was designed by Rune Elmqvist (1906-1996) and Ake Senning (1915-2000) of Sweden. It was implanted into a patient named Arne Larsson (d2001) in 1958 but failed after three hours. Fortunately, Larsson survived. In fact, he was almost indestructible, outliving 22 other pacemakers before he finally expired in 2001.

The first internal pacemaker with a reasonably long life came into existence as a result of an electrical error. During the late 1950s, a professor at the University of Buffalo named Wilson Greatbatch decided he needed to install a 10,000 ohm resistor into a machine he was developing to record human heartbeats. Confused by the similar colour-coding, he fitted a 1,000,000 ohm resistor by mistake. When he tested the machine he heard a strong pulse, then a silence of about a second, then another pulse. Greatbatch realised that he had inadvertently constructed a pacemaker. It took him two years to shrink the device to implantable proportions. He patented it on 22 July 1960, since when pacemakers based on Greatbatch's design have prolonged millions of lives.

PAINT-BY-NUMBERS

Leonardo da Vinci (1452-1519) contributed to the paint-by-numbers craze that took over the US in the early 1950s. Artist Dan Robbins remembered reading that when da Vinci was overwhelmed with large, complicated commissions, he would ask apprentices to block in areas of the canvas by giving them numbered patterns to follow. He finished the paintings himself, but it was a clever way to instruct pupils and save time all in one. While thinking about this story, an idea dawned on Robbins: why not do patterns for paintings that people can finish for themselves? He set about developing paint-by-numbers kits consisting of two brushes and up to 90 pre-mixed paints ready to be applied to numbered

DID YOU KNOW?

- The original kits sold by Palmer Paint Company of Detroit, Michigan, coined the phrase, 'Every Man a Rembrandt!'
- 'By the numbers' temporarily replaced the popular expression 'by the book' when referring pejoratively to formulaic consumer products of the time.
- Pop Art took on paint-by-numbers as a commentary on popular culture in the 1970s. Andy Warhol (1928-1987) purchased one of five abstract paint-by-number kits created by artist Paul Bridgewater in 1978.
- Many still collect the original paint-by-numbers kits from the 1950s, spotting vintage samples at garage sales, flea markets and internet auction sites.

spaces onto an accompanying board or canvas. As the spaces were filled in, the picture gradually revealed itself.

Paint-by-numbers proved so popular that its appeal reached as far as the White House. In 1954, presidential appointment secretary Thomas Edwin Stephens distributed kits to staff and friends of the Oval Office. Assuming that President Eisenhower (1890-1969) himself expected them to rise to the occasion, many of the recipients dutifully completed their kits. Before long, Stephens had himself a gallery of paint-by-numbers, which he eventually mounted in the West Wing corridor.

The paint-by-numbers fad reached every class level of post-war America, during an era in which free time, new homes and disposable income abounded. However, it was not universally liked. Many saw it as a testament to the mindless conformity plaguing the American population of the 1950s. Others, who would never have picked up a paintbrush without the security of completing a finished product, found it liberating to engage in some kind of artistic activity. One enthusist wrote to the magazine *American Artist*: 'Why oh why didn't you or someone else tell me before how much fun it is to use these wonderful paint-by-number sets?' The editor's response was not recorded.

PAJAMAS/PYJAMAS

'Pajama' is an Urdu word meaning 'leg clothes'. In the garments' native India, they are worn during the day rather than as nightwear. Pajamas were brought to Britain during the colonial era and became fashionable items for men to wear in bed during the 1880s. The British were not, however, the first people to use them in this manner. François Pyrard, a Frenchman held captive in Goa between 1608 and 1609, mentioned in his memoirs that the Portuguese residents of the city always wore pajamas to bed.

PANTYHOSE

Tights or pantyhose are thought to be derived from the 'hose' worn by European men in the Middle Ages. These were two singular knitted leg-coverings, fastened across the back, then across the front, with a space left open for hygiene habits.

In England in the late 1500s, William Lee (c1550-1610) invented the mechanical knitting machine, and by the twentieth century, hosiery was being produced in cottons, wools and silks. These fibres were problematic, however, as they had no elasticity. This meant that many different sizes still had to be made, and the material tended to stretch and sag. It wasn't until the

Pantyhose are thought to come from the hose worn by men in the Middle Ages, such as this fifteenth-century German patrician

DID YOU KNOW?

- The difference between pantyhose and tights is all in the denier (defined as the mass in grams per 9,000 metres). It is dependent on the weight of yarn used and the thickness of the knit. So, anything up to 40 denier is referred to as pantyhose and anything over as tights.
- Pantyhose fetishism is now the third most popular clothing fetish item in the western world, and more popular than stockings. This may seem odd as it is commonly assumed that the stockings and suspenders combination is sexier... but apparently not.

invention of nylon that pantyhose really became popular. This was patented in 1937, when Wallace Carothers (1896-1937), a research chemist for DuPont, in the States, discovered that a mix of coal, tar air and water produced a strong and flexible fibre. When these liquid polymers were blown through ultra thin nozzles, they immediately hardened to form sturdy fibres, thinner than human hair. The finished result presented fibres that were as sheer as silk but stronger than cotton or wool.

The first day of sale for DuPont's nylon stockings in May 1940, sold a whopping 72,000 pairs. Production came to a halt during World War II when nylon was used for war supplies only, and women were forced to go bare legged. Times had become so desperate that by 1945, with the first post-war sale of hosiery in San Francisco, 10,000 shoppers went on a rampage in their desperation for new pantyhose.

The next major improvement came in 1959 with the creation of Lycra® (also known as Spandex). Essentially a texturisation of yarn under heat treatment, it meant that hosiery could be stretched by more than 600% without breaking, and would always recover its original shape. Also, using Lycra meant there were no problems with static cling or piling. This development completely revolutionised the hosiery industry.

PARACHUTE

The concept of the parachute emerged almost three centuries before the reality. Around 1485, Leonardo da Vinci (1452-1519) made a sketch of a man falling through the air attached to a pyramid-shaped device that is unmistakably a prototype parachute. He accompanied the drawing with the caption: 'If a man has a tent roof of caulked linen 12 braccia [arm's lengths] broad and 12 braccia high, he will be able to let himself fall from any great height without danger to himself.' There is, however, no more reason to suppose that Leonardo actually tested the device than to think he buzzed around in a real-life version of another of his drawings, the helicopter.

Predecessor to the parachute? Louis Sebastian Lenormand developed his first parachute out of two umbrellas, using it to safely jump from a tree. Two years later he moved onto tests with animals.

In 1783, a French watchmaker named Louis Sebastian Lenormand unveiled a piece of safety equipment designed to allow people to escape from burning buildings, which he called a 'parachute'. Lenormand demonstrated its fall-breaking abilities by dropping several hapless animals from the roof of the Montpellier Observatory. It took 14 years for a human being to summon up the courage to take the plunge. The first parachutist was André-Jacques Garnerin (1769-1823), described in one contemporary account as 'a small, peppery man with a spade beard and waxed moustaches'. On 22 October 1797, Garnerin demonstrated the new art over the Parc Monceau in Paris. His parachute consisted of a basket connected by a long wooden pole to a rigid canopy about 23 feet in diameter. The canopy was in turn suspended from the balloon that bore him aloft. When Garnerin reached an altitude of approximately 3,000 feet, he pulled the release cord and initiated a terrifying descent. The parachute did indeed slow his fall, but it lurched violently, swinging the basket around like a pendulum. One several occasions, Garnerin was flung higher than the canopy itself. When he returned to Earth, he was violently sick. The stability problem was only solved in 1804, when an astronomer named Joseph-Jérôme le François de Lalande (1732-1807) came up with the idea of cutting a small hole in the apex of a parachute. This reduced the air pressure beneath the canopy and eliminated the alarming oscillations.

PARKING METER

The man we have to thank or curse for the invention of the parking meter was a journalist named Carlton C Magee. In 1927, Magee moved to Oklahoma City from New Mexico to found the *Oklahoma News*. Between

US Patent 2,118,318 24 May 1938

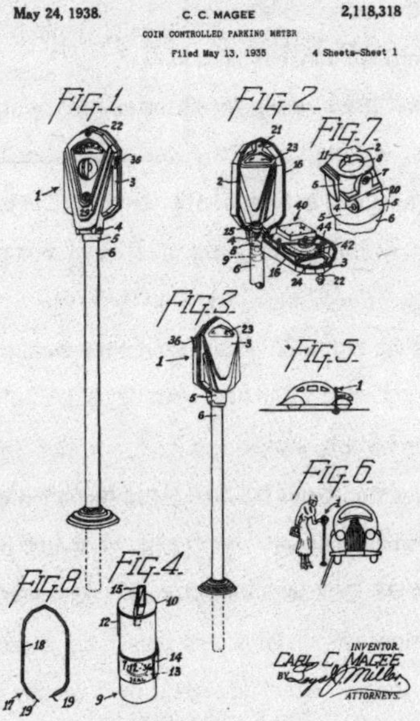

The start of so much misery:
Carlton Magee's original parking meter.

1913 and 1930, the number of cars in the state had multiplied from 3,000 to half a million, giving the city the most appalling traffic problems. Shoppers were unable to find anywhere to park because downtown workers were bagging all the available parking spaces early in the day, and commerce was suffering as a result.

In 1932, Magee was made chairman of the Oklahoma City Chamber of Commerce Traffic Committee, and he filed the first patent for a parking meter in the same year. The first parking meters went into operation in Oklahoma City in 1935.

The devices were installed experimentally in Britain in 1955 before being adopted permanently by Westminster City Council in 1958. The 625 meters installed charged half a shilling for an hour's parking and there was a fixed penalty of £2 for illegal use. Two years later,the first traffic wardens appeared on the streets of New York and London.

PARTY BALLOONS

The first balloons used purely for entertainment purposes were made from inflated animal bladders. They were important tools-of-the-trade for medieval jesters and are frequently referred to in later literature. The novel *Swiss Family Robinson*, for example, which was published in 1813, contains the line: '"Papa," said Jack, "can't you make me a balloon with this piece of whale entrail?".' Rubber balloons were developed in 1824 by the British scientist Michael Faraday (1791-1867) for use in experiments involving hydrogen.

The following year, similar balloons were marketed as toys by another Briton, Thomas Hancock. Purchasers received bottles of rubber solution with which to make their own balloons and small syringes with which to inflate them.

PASSPORT

One of the earliest references to a passport occurs in the Old Testament. In chapter two of the book of Nehemiah, which was written around

400BC, the eponymous hero asks the Persian King Artaxerxes to allow him to return home to rebuild the walls of Jersusalem. 'If it pleases the King,' he says, 'let letters be given to me to give to the governours of the province beyond the river, that they may let me pass through until I come to Judah.' Similar letters guaranteeing safe passage through hostile territory have been around for millennia, but passports only became mandatory for ordinary travellers around the time of World War I. The international system was formalised at a conference at the League of Nations in 1920.

PAYPHONE

As more and more people acquire mobile telephones, the payphone is becoming a threatened species, but when the device was introduced, it revolutionised communication. The first coin-operated telephone was designed by William Gray and installed in a bank in Hartford, Connecticut in 1889. (Gray's other major inventive contribution was the inflatable baseball chest protector.) It was a 'post-pay' machine, meaning that the caller was required to deposit coins at the end of his or her conversation. The more familiar 'pre-pay' public telephone made its debut in Chicago in 1898.

PENCIL

In Ancient Rome, scribes wrote with thin metal sticks called 'styluses'. They were often made from lead, which left a distinct mark when rubbed against papyrus. As a consequence, the cores of pencils are still known as leads, even though they have been made from graphite for several centuries, although

Romans wrote with metal sticks called 'styluses'. These were often made of lead, which is why pencil cores are known as leads, despite now being made from graphite.

another possible explanation is that scientists originally thought graphite was a form of lead. The breakthrough came when a storm exposed a seam of high quality graphite in Borrowdale in the English Lake District in about 1500. The local shepherds found that the material was very useful for marking sheep. By 1558, the first pencils were being manufactured in the Cumbrian town of Keswick. Although inferior graphite pencils were made in Nuremberg in Germany from about 1662, the Borrowdale region enjoyed a virtual monopoly on their production until 1795. In that year, a French chemist named Nicholas Conté devised a method of making pencil leads in which powdered graphite was blended with clay. The mixture was then shaped into thin sticks which were hardened in furnaces.

PHOTOCOPIER

Before the invention of photocopiers, technical drawings had to be copied out by hand. In the case of Chester F Carlson (1906-1968), a Seattle-born physicist turned patent clerk, this tedious work was made considerably harder by arthritis and short-sightedness. Carlson decided there had to be an easier way. Using his knowledge of photo-conductivity, the process by which certain chemicals change their electrical charges when exposed to light, he set to work in the kitchen of his apartment in Jackson Heights in the New York borough of Queens. On 22 October 1938 he found the answer he had been looking for. Carlson had his assistant Otto write '10-22-38 Astoria' on a glass microscope slide and waited for the ink to dry. Then he pulled the kitchen blinds down and laid the slide on a sulphur-coated zinc plate which had been vigorously rubbed to give it an electrical charge. He switched on an incandescent light for a few seconds, removed the slide from the plate and sprinkled some lycopodium powder on the space it had vacated. When he blew the powder away, he saw a near-perfect reproduction of the message '10-22-38 Astoria' on the sulphur surface. At that moment, as Carlson later put it: 'I knew I had a very big tiger by the tail.' Nevertheless, it took him several years to find a business willing to invest in the invention. He was rejected by IBM and General Electric among others before finally licensing the copying process to the Haloid Company of Rochester, New York in 1947. The name 'Xerox', derived from the Greek for 'dry writing', was registered as a trademark the following year. The company's first commercial photocopying

US Patent 2,297,691 6 October 1942

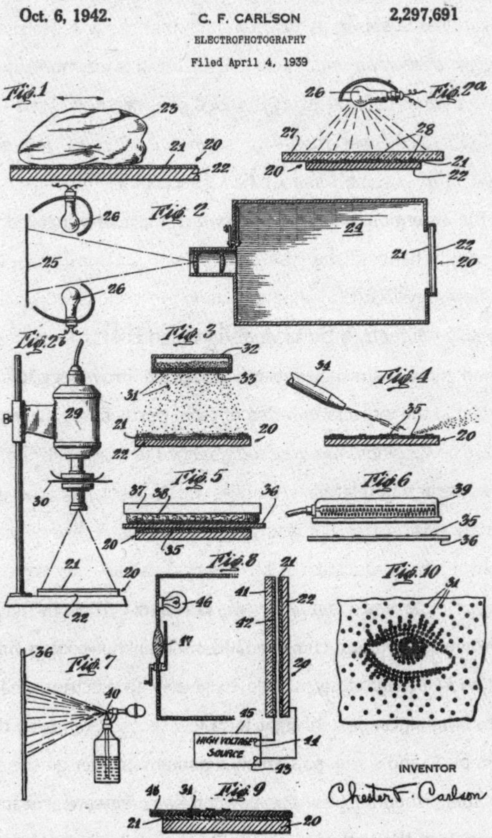

Chester Carlson's 1942 photocopier patent design.

machine, called the Xerox 914, went on sale in 1959. It was so successful that the Haloid Company changed its name to Xerox in 1961.

PIGGY BANK

Strange as it may seem, the origin of the term 'piggy bank' has nothing to do with hogs. During the Middle Ages, Western Europeans used to keep household essentials such as salt in jars made from a cheap orange clay known as 'pygg'. (The term may be linked etymologically with the low quality metal known as 'pig iron'.) At some point in the eighteenth century, an anonymous ceramicist decided to make a money jar in the shape of a pig as a visual pun. An added consideration is the fact that the household pig was crucial to survival in days of yore and was slaughtered in autumn when it had been sufficiently fattened with scraps. The analogy with piggy banks, which must be destroyed to be of use, is hard to ignore.

PINBALL MACHINE

Pinball's humble beginnings couldn't be further from its gaudy image. Originating in the respectable milieu of table games that grew out of field games like bowls, pinball is most directly related to a descendent of billiards called bagatelle, which in the 1700s was widely popular, but can now only be found in parts of Wales and northwest England. There were many variations in rules and table size, but the basic set up consisted of a table eight feet long and two feet wide, with one semi-circular end. Inside that semi-circle would be eight numbered holes and a ninth in the centre, like an imaginary eight-point compass. Balls would then have to be potted into the holes from the square end using a cue stick.

Bagatelle became very popular in America after it was imported by French soldiers fighting in the War of Independence; but it was a Cincinnati-based Brit called Montague Redgrave who, in 1871, decided to replace the cue with a spring-loaded handle and make other crucial changes such as compacting the game's size and replacing the balls with

A bagatelle table. Note the introduction of the pull-back spring at the bottom right of the table, and the lane for the ball to run around the top – both familiar features from the modern pinball table.

marbles. In 1889, Louis Glass and William S Arnold invented a coin mechanism and added a glass cover – to deter cheats – and the game became a familiar sight in pool halls across the US.

Enterprising parts dealer Leo Berman developed his own version of the game on the threshold of the Great Depression, angling the playing surface and surrounding the holes with pins to increase the challenge. This did the rounds for a while until it was outmanoeuvred by David Gottlieb's more showbizzy 'Baffle Ball' and 'Bally's Ballyhoo' games, which came to be smash hits in a time of crippling economic strife. From this point, pinball rapidly evolved into how we know it today, starting with the modifications of Harry Williams. As well as adding flippers to the board and a power-cutting anti-tilt trigger, Williams used electromagnets to propel the new, steel balls out of the scoring holes, and a surface-mounted solenoid that forced the ball to travel in unexpected directions. Joining forces with manufacturer Fred McCellam, he promoted a machine called Contact under the company name Pacific Amusements Co. Although pinball production suffered during World War II, the post-war era was a golden age that brought kickout holes, bumpers, lights, automatic ball-lifts, multiple player functions and electromechanics to the established framework.

In 1969, the game attracted the ultimate tribute when The Who released their concept-album, 'Tommy', a bizarre story about a deaf, dumb and blind kid who, despite his handicaps, 'Sure plays a mean pinball'. The signature track, 'Pinball Wizard', bolstered the game's popularity, but when interviewed about his skills at the table, frontman Roger Daltrey (b1944) admitted that he was, in fact, 'useless'. By the 1980s technology had moved on and the pinball business declined, although it did make a partial comeback in the 1990s, when the Pinball Hall of Fame opened in Las Vegas, displaying working models from the 1950s onwards. Despite this, there is only one pinball manufacturer left in the world today – the Stern Pinball Company – which requires every member of staff to play pinball for at least 15 minutes a day.

PIZZA

Be careful who you mention this to, particularly when in the presence of Neapolitans, but the pizza was probably invented by the Greeks rather than the Italians. The Ancient Greeks ate a flat, round bread baked with various herbs and spices called plankuntos. If, despite our advice, you do

Naples, widely considered the birthplace of pizza, though in fact the Ancient Greeks ate an early variant called plankuntos first. However they did introduce it to Naples, which was a colony of theirs, and it was almost certainly here that it acquired its trademark tomato paste topping and developed into the dish we know today.

manage to offend an Italian, try mollifying them by pointing out that Naples was originally a Greek colony (the city's name derives from *neo polis*, meaning 'new town'). On the other hand, this might make matters worse. If all else fails, draw their attention to the fact that the Neapolitans were almost certainly the first people to think of adding tomato paste to the basic recipe. By the eighteenth century, street vendors were walking around the city selling pizzas from small tin ovens mounted on their heads. The first true pizzeria opened in 1830 at 18 Via Port'Alba in Naples. Named Antica Pizzeria Port'Alba, it is still doing thriving business.

The pizza became a national and then international phenomenon after a chef named Rafaele Esposito served one to King Umberto I (1844-1900) and Queen Margherita (1851-1921) during a royal trip to Naples in 1889. To sound an appropriately patriotic note, Esposito made the pizza in the colours of the Italian flag, using tomatoes, mozzarella and basil. He named it a 'margherita' in honour of the queen.

PLASTICINE

Plasticine was invented by art teacher William Harbutt (1844-1921), who developed the substance in 1897 for his students for modelling purposes. It was patented in 1899 and Harbutt set up a factory in Bathampton at the turn of the century. His malleable gunk proved to be an instant hit with children, but less so with parents, for whom it became a carpet-clogging nightmare. An early example of a Plasticine-animated film was *A Sculptor's Welsh Rarebit Nightmare* (1908) but this medium remained largely experimental until the appearance of Plasticine figure Morph in *Take Hart* in the 1980s, and in 1980s films such as *Meet the Raisins* and TV shows like *Trap Door.*

Plasticine is made from aliphatic acids, calcium salts and petroleum jelly. At room temperature it remains flexible but firm and can be reused umpteen times. In fact, the further away it is kept from heat, the better, as the Harbutt factory discovered to its cost when it burned to the ground in 1963. The downside to plasticine's reusability is that when its colours have been mixed together too often, it goes permanently brown.

In 1956, a rival substance called Play-Doh was created, with a recipe that, even today, remains a closely guarded secret. This sweet-smelling, non-toxic putty had the ability to dry up and set when exposed to the air for long periods, unlike its predecessor.

Plasticine reached its highest artistic watermark in 1986, when Aardman Animations collaborated with the Brothers Quay and director Stephen Johnson on Peter Gabriel's legendary 'Sledgehammer' music video. This deliberate pulverisation of Gabriel's slick image blew the roof off the animation community and won a slew of industry awards.

Aardman Animations went from strength to strength over the next decade or two thanks largely to its Wallace and Gromit characters, who were showered with BAFTAs and Oscars. Aardman suffered its own serious warehouse fire, in October 2005, but continued undaunted.

William Harbutt died in New York in 1921. After the fire of 1963, the factory was rebuilt and his family kept the business running until 1983.

The name 'Plasticine' lives on as a trademark.

PLAYING CARDS

The earliest reported use of playing cards in Europe was in the late 1300s, when they were imported from the Islamic empire; before that they first existed in China, the home of paper. These early playing cards bear little resemblance to their modern-day counterparts, but were more like dominoes: slender, ornate, and more or less book-mark shaped. In fact, it is thought that sometime in the tenth century, the Chinese may have initially invented cards as 'paper dominoes' in order to use them in new games. At first, the packs were limited to only two suits – sticks and coins – and the suits of cups and swords are thought to have been added in the Moslem world, along with court cards, though these would not have literally represented a royal personage until reaching Europe.

The shape of playing cards has also varied; there are examples of circular cards (still used in India, with a mind-boggling eight to 10 suits), which look a bit like decorative beer mats. The European card we know today was

Originally, playing cards had two suits: sticks and coins, which were later joined by cups and swords. Even today variations exist: Germany has acorns, leaves (both top left), bells and hearts; Spain has coins, cups (both top right), swords and clubs – but no queen cards at all.

perfected in France in the fourteenth century, when cards could be produced on a large scale for the first time, and were affordable for the general public. It was also in France that the suits of hearts, spades, diamonds and clubs were settled on, although different suits are still commonplace in other European countries; Germany has hearts, leaves, bells and acorns; Spain uses the more archaic coins, cups, swords and clubs.

Over the next 200 years, the design of playing cards evolved differently in each country. Germany and Switzerland, for example, lost the Aces from their packs; you will not find a queen in Italian or Spanish packs; and the joker was born in America in the late 1800s. Such features as standardisation of size and the addition of corner indices were applied fairly ubiquitously due to their undeniable usefulness.

As printing technology continued to evolve, so too did playing card design, and artistic or themed packs became increasingly prevalent at the start of the nineteenth century. At the same time, an offshoot of the conventional pack had also established itself– the Tarot, or 'trump' pack, with an additional 22 unsuited cards, including characters such as The

Hanged Man and Death, often represented by detailed and sinister illustrations. Contrary to popular belief, these trump cards were added to enliven card gambling; their relationship with fortune-telling and the occult did not develop until the mid-1700s, when Tarot packs had already been in use for some 200 years. It was long thought that tarot cards originated in ancient Egypt, but this assumption has now been largely discredited, despite what the fortune-tellers say.

POCKET CALCULATOR

The first incarnation of a counting machine was the abacus: a framed rack of rods and sliding counters, designed to represent a range of decimal columns and help with totting up figures. Thought to have developed from an ancient Babylonian system involving sand in a tray that was used for simple finger-writing, the abacus first sprang up in Egypt around 500BC and later emerged in second century China. From the mid-1300s, it was immensely popular there. Other, mechanised counting devices followed, like John Napier's (1550-1617) Rabdologia (1617), which became known as 'Napier's Bones' as it included a set of spooky numbered rods that were used to perform calculations. Echoes of bone-reading witchery accrued around Napier's machine, and earlier rumours of its creator's devil-worship couldn't have helped. William Oughtred (1575-1660) invented the somewhat less creepy slide-rule in 1622, and Blaise Pascal (1623-1662) created an adding machine made up of nine gear wheels in 1642.

In the twentieth century, two things swung into play that would help arithmetic machines to work faster and be made smaller: the transistor (invented 1947), and the integrated circuit, aka the microchip (invented 1958). All that was required was a catalyst, which in this case was Texas Instruments. It had already enjoyed enormous success with the transistor radio, and was looking to consolidate its position as pre-eminent tech manufacturer for the mass market. Company president, Patrick Haggerty, saw an opening for a calculating machine with similar appeal, and put Jack Kilby (1923-2005) on the case, the co-creator of the integrated circuit. Kilby

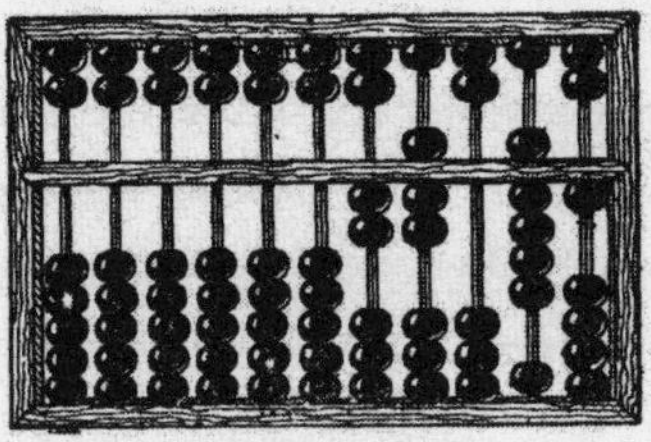

Calculators old...

US Patent 3,819,921 25 June 1974

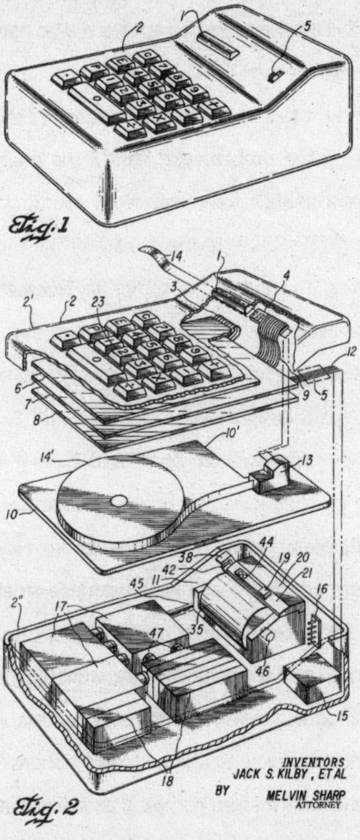

... and new(ish). Kilby's pocket calculator patent.

assembled a team that set about trimming the model size of the design and cutting as many cost corners as possible to meet Haggerty's specification of a US$100 retail price. Kilby was so good that, even though he filed for a calculator patent in 1967, the manufacturing technology to produce it wouldn't arrive for another four years. Once it had been released, it sold five million units in 1972. And, at a hundred bucks a pop, well... you do the maths.

POGO STICK

The pogo stick was conceived in 1919 by German George Hansburg, who fashioned it out of wood and intended it be a diversion for children, and a moneyspinner for himself. It was not long before it attracted the attention of the American Gimbel Brothers department store, who approached Hansburg to replicate his design for the US market. Hansburg shipped out a crate load of samples for the firm to test out, but unfortunately they'd warped in the damp cargo hold by the time they'd crossed the Atlantic. Gimbel asked Hansburg to upgrade his stick, so he replaced the wood with metal.

That done, the 1920s became the toy's heyday, thanks in part to an unlikely project that Hansburg undertook – to teach the Ziegfield Follies how to pogo. His efforts paid off handsomely and the first entirely pogo-mounted marriage took place onstage with the Follies in the same decade. The pogo stick's basic design went unchanged till Gordon Spitzmesser invented the gas-powered stick in 1960. Essentially it was a pogo with a petrol motor; the gasoline kicked in on each bounce and sent you higher with each pogo. Its popularity was short-lived.

Spitzmesser's son, Edwin, remembers his father's invention thus: 'I would test it each time he tinkered with it. It ran on butane, but really jumped high when we once used a mixture of acetylene and oxygen. He used a model 'A' Ford spark coil, which delivered quite a few volts to the unprotected spark coil plug – as I once found out when my zipper came into contact with it briefly. I haven't had a thrill like that since.'

The world record for endurance pogoing is held by Gary Stewart, of Huntington Beach, California. On 25 May 1990, he jumped 177,737 consecutive times for 20 hours and 20 minutes. Touchingly, his feat was the summit of long-held ambition: 'Ever since I was in the first grade I would get a Guinness book for Christmas every year,' he said at the time. 'Getting my name in it was... kind of a childhood dream.' The record held for the longest distance travelled in a single journey on a pogo stick is 23.11 miles, accomplished by Ashrita Furman in New York in 1997.

POPCORN

The pre-colonial inhabitants of America used three kinds of corn: fresh sweet corn for immediate eating, field corn for animal feed and so-called Indian Corn for popping. The latter had a high enough water content to inflate when it was heated. In 1948 and 1950, ears of popcorn estimated to be 4,000 years old were found in Bat Cave in New Mexico. They ranged in size from half an inch to two inches. The Amerindians didn't just eat popcorn. They used it for divination purposes, drawing messages from the direction in which kernels thrown on the fire jumped when they popped, and also made it into jewellery. When Cortés invaded Mexico in 1519, he found the Aztecs constructing elaborate necklaces and headdresses from popcorn and draping them over statues of their gods. Columbus also came across popcorn when indigenous tribes tried to sell it to his men. The first popcorn-making machine was an adapted peanut roaster built by an American called Charles Cretors in 1885.

Popcorn wasn't always just for eating: the Aztecs made necklaces and headdresses of popcorn and draped them over statues of their gods.

POSTAGE STAMP

The world's first adhesive postage stamp, the 'Penny Black', was issued in Britain on 6 May 1840. It was decorated with a profile of the young Queen

Victoria wearing a short, high ponytail and a tiara. The stamp was the brainchild of a progressive schoolteacher named Rowland Hill (1795-1879). Hitherto, postal fees had been payable by the recipients of letters rather than the senders. This arrangement had caused all sorts of problems. Unwilling addressees had often refused important letters and the impecunious had been in no position to pay for them. Meanwhile, unscrupulous customers had taken to conveying information for free by writing coded messages on their envelopes. The recipients could then read them and decline to cough up.

Hill was supposedly moved to action by the sight of a tearful young woman unable to pay for a letter from her fiancé. In 1837, he published a pamphlet entitled 'Post Office Reform: Its Importance and Practicability' which called for an end to the system of payment on receipt. The author suggested two alternatives. The first was a pre-paid letter form and the second was the adhesive stamp. In 1839, the government decided to adopt Hill's suggestions and gave him a two year contract to implement them. The letter forms were not a success – stationery manufacturers felt threatened by them and orchestrated a negative campaign – but the stamps were a triumph.

PRESSURE COOKER

If water is heated under pressure, its boiling point rises beyond 100°C/212°F. This was the principle behind the pressure cooker or 'steam digester' invented by the Frenchman Denis Papin (1647-c1712) in 1679. His day job had prepared him admirably for the task: Papin was the assistant of the British physicist Robert Boyle (1627-1691), best known for formulating the law that governs the relationship between the volume of a gas and its pressure. He later worked for Robert Hooke at the Royal Society in London. On 12 April 1682, Papin used his creation to prepare a meal for the members of the august society. The diarist John Evelyn (1620-1706) described the dinner as 'all dress'd, both fish and flesh, in Monsieur Papin's Digestors, by which the hardest bones of beefe itself, and mutton, were made as soft as cheese without water or other liquor.' Papin was prudent enough to include a pressure valve in the lid of his cooker; had he neglected to do so, it would have effectively been a bomb. Despite the excellence of the meal, Papin's invention was originally exclusively used in industry. The first commercial pressure cooker made its debut in 1939.

PRETZELS

The treat that nearly finished off George W Bush has an ancient pedigree. 'Pretzel' is derived from the Latin *pretiole*, which means 'little gift'. According to legend, the snack was invented by an Italian monk in the early seventh century and given to children who did well at their lessons. The shape of the pretzel can certainly be seen as loaded with Christian symbolism. The central bars are said to represent the arms of a praying child and the three holes to represent the Trinity. In time, the 'pretiole' travelled over the Alps to Germany and became the 'bretzel' or pretzel.

PROZAC®

Fluoxetine hydrochloride, or Prozac, was developed during the 1980s by a group of researchers at the Eli Lilly pharmaceutical company headed by Klaus Schmeigel (b1939) and Bryan Molloy (1939-2004). Prophetically, Schmeigel, one of the most influential biochemists of modern times, was born in a German town called Chemitz. Prozac was one of a class of substances with the intimidating name 'aryloxyphenylpropylamines' patented by the team in 1982. The drug proved remarkably effective at alleviating depression, which it did by raising levels of the key neurotransmitter serotonin in the brain.

The FDA approved Prozac for the treatment of depression at the end of 1987. It was quickly proclaimed a miracle drug and received invaluable publicity in March 1990 when *Newsweek* magazine ran a cover story entitled 'The Promise of Prozac'. Since then, it has been prescribed to more than 50 million people, making it the most popular anti-depressant in the world.

PSYCHOANALYSIS

The first recipient of psychoanalysis was a 21-year-old woman named Bertha Pappenheim (1859-1936), or Anna O as she was referred to in her case notes. Although it was her case that prompted Sigmund Freud (1856-1939) to develop his fully fledged psychoanalytical theory, she was actually treated by his friend and fellow Viennese psychiatrist Josef Breuer (1842-1925). When Pappenheim first visited Breuer in 1880, she was suffering from seizures and paralysis in one arm. When she talked about the origin of her symptoms, Breuer noticed that she frequently remembered a previously repressed episode and experienced an outburst of emotion. Afterwards, the relevant symptom would improve or disappear. In one session, for example, at a time when Anna/Bertha was refusing to drink, she recalled seeing a woman drink from a glass previously used by a dog. When she had experienced the disgust associated with the incident, she promptly asked for a drink of water. Breuer named this phenomenon 'catharsis', the Greek word for 'cleansing'. Thirteen years after she finished her treatment, Anna O's case became the focus of Freud's seminal book *Studies in Hysteria* (1895), co-authored with Breuer. Although she later spend time in a sanatorium, Pappenheim emerged to enjoy a productive adulthood as a pioneering social worker.

Sigmund Freud, generally considered the father of psychoanalysis, although it was his friend Josef Breuer who treated the first psychoanalysis patient, Bertha Pappernheim.

PVC

Polyvinyl chloride was first synthesised in Germany in 1838 by Henri Victor Regnault (1810-1878) and then again by Eugen Baumann (1846-

1896) in 1872. It is a plastic commonly known as PVC and its discovery falls into the category dubbed 'accidental'. The story goes that Regnault mistakenly exposed some vinyl chloride to sunlight and noted that some white powder formed on it. Over 30 years later – just a year after Regnault's 1871 retirement – Baumann rediscovered the discovery. He found that the white powder had unique properties, in that it was unaffected by solvents or acids.

In the 1920s the substance was still thought to be 'worthless' until a young scientist named Waldo Semon (1898-1999) came along. He was supposed to be trying to find a way of coating rubber onto metal for the American tyre giant, BF Goodrich; but when his rubber supplies ran out one day, he started experimenting with polymers and 'accidentally' (his words) found a way to make PVC more elastic. He put the initial, metal-coating project on the back burner and concentrated on developing PVC. A new industrial material was born: malleable, durable and yet not adhesive... in other words, a goldmine.

However, BF Goodrich almost fatally underestimated the new polymer and came close to canning production, as they made scant return on the PVC foot soles they released. Semon saved the day, pitching to his vice-president the idea of coating fabrics with PVC to make them waterproof. The VP – a camping enthusiast – jumped at the proposal, and the company made a mint from treated textiles. Semon then went back to his original project, coming up with over a hundred ways of bonding rubber and metal. By the time he died, at the age of 100, he had 116 patents under his belt and a place in the Inventors Hall of Fame.

First commercially synthesised in the 1930s, PVC is now the second most-used plastic, just pipped by polyethylene, and over 25 million metric tons of it are produced every year. PVC is incredibly useful as it is inflammable, un-reactive and inexpensive to produce. Over 50% of PVC is used in construction materials like electrical wire insulation and pipes, along with modern window frames that are such a common sight in today's double-glazed suburbs. There's even PVC clothing... if you're that way inclined.

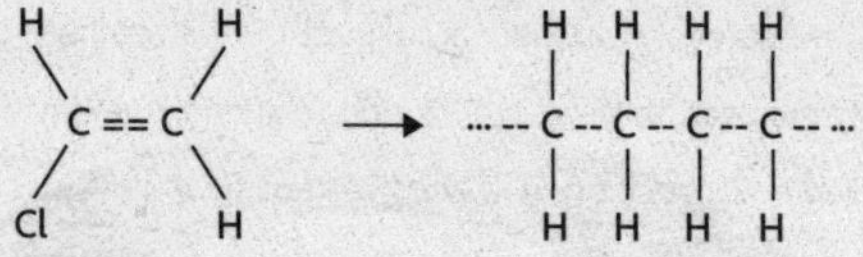

Vinyl chloride monomer — Polyvinyl chloride polymer

Chemical composition of PVC

Unfortunately, its benefits are offset by certain negative effects on health and the environment: it contains carcinogenic chemicals, is inefficient to recycle and releases harmful by-products. These include hydrochloric acid, which contributes to acid rain.

Q

Q-TIPS®

It is probably just as well that in 1926 the creator of the cotton swabs that are used to clean infants' ears decided to change their name to Q-Tips (the Q stands for 'Quality'). They were initially called 'Baby Gays'. They were invented in the early 1920s by a Polish-American named Leo Gerstenzang. He got the idea from his wife, who made delicate swabs for their baby daughter by wrapping cotton wool around the tips of toothpicks. Sensing a market for the implements, the couple founded the Leo Gerstenzang Infant Novelty Co. in 1923. Leo developed a machine that manufactured similar items, with the crucial difference that they were, as the advertising boasted, 'untouched by human hands'.

R

RADIATOR

The concept of central heating dates back to the Romans and their Hypocausts. This was an impressive system of under-floor tunnels which circulated hot air heated by furnaces to heat the floor above.

However, we were a little slower to develop an efficient modern day equivalent. James Watt (1736-1819) (the man who invented the steam engine and had the unit of Watt named after him) piped steam into his factories to heat them in 1784 and patented the method in 1791. But it was expensive and brought up dirt and soot along with the steam.

In the nineteenth century, Angier Perkins, continuing his father's work, developed the first central heating to be found in common usage. Pressure circulated hot water from a boiler to a closed system of pipes that coiled in places on the wall, like a radiator. The first building to be heated in this way was a greenhouse in Fulham, London in 1832. The system was effective but had its flaws – the heat from the pipes nearest the boiler was far greater than that of the pipes furthest away and the heating either had to be on or off throughout the whole building.

A Roman Hypocaust, history's earliest example of central heating. Hot air from a furnace was circulated in a system of under-floor tunnels, which then heated the floor above. The concept of under-floor heating has now returned in today's ultra-modern apartments.

In 1837, a man named Joseph Nason travelled to England from America to meet with Perkins and together they designed valves that meant that the water could be tapped off or on in each room. Nason installed a coiled, valved, hot water radiator into a counting room of the Middlesex Mill in Massachusetts in 1841.

Some say the modern radiator was invented by the American William Baldwin the early twentieth century. However, the people of Samara, Russia, claim its origins are located in St Petersburg, invented by a German man named Franz San-Galli in 1855. The structure was similar to the radiators of today, if a little more ornate, and worked in the same way.

Ironically, radiators work by convection and not radiation. Some say Nason is responsible for christening it with this misleading name, when he and a man called Briggs brought out their radiator in 1863. San-Galli called his invention the 'hot box' which seems much more appropriate.

Today, it seems we are returning to the concept of our Roman ancestors, with the advent of electric under-floor heating, producing a much more energy efficient way of heating a room... by radiation.

RADIO

The development of radio would not have been possible without the earlier development of telegraphy (from the Greek, meaning to 'write with/over distance'), pioneered by Claude Chappe (1763-1805) in 1792 with his invention of the optical telegraph machine. In 1838, Samuel Finley Breese Morse (1791-1872), probably better known for the code he invented, patented his electric telegraph machine and, in 1844, sent the first telegram from Washington DC to Baltimore, Maryland. Morse's message read 'What hath God Wrought'.

Another notable experiment involving wireless messaging was undertaken by Mahlon Loomis (1826-1886), a dentist from Washington DC, who successfully sent the first wireless signals in 1865. By setting up two kites 18 miles apart and earthing them with closed copper wires, he was able to transmit signals using the Earth as conductor. But it wasn't until

US Patent 645,576 20 March 1900

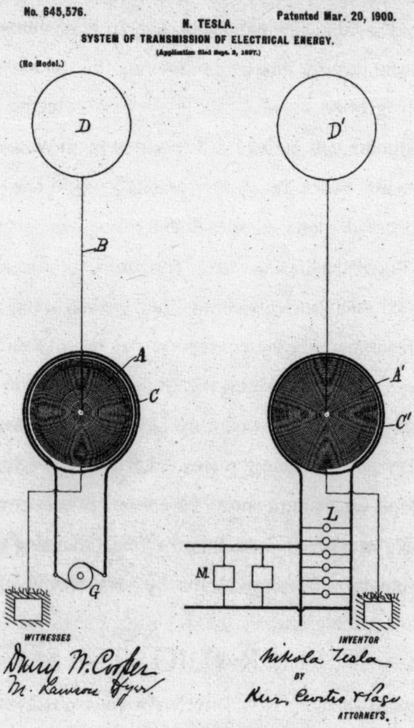

Nikola Tesla now holds the first patent for the radio, although it was taken off him during his lifetime and only given back posthumously – in part because of his eccentric claims to have invented a 'death ray' and received extraterrestrial signals from Mars.

later, in the 1880s, that Heinrich Hertz detected the first radio waves. Interestingly, he was actually testing the data of another scientist when he accidentally set up a radio transmitter and receiver from copper wire coiled in jars, and observed sparks bridging the gap between the separate wires.

However, the invention of the first radio must be attributed to three men, as it is unclear (and still hotly contested) which scientist in reality got there first. The most mysterious and perhaps first transmission was ostensibly carried out by Nathan Stubblefield (1860-1928), a Kentucky melon farmer who claimed to have sent the first human voice radio message, 'Hello Rainey', to a field partner in 1892. However Stubblefield was something of an eccentric and extremely protective of his ideas – there are rumours that during his demonstration in Washington DC that year, a 'radio-like' device was stolen from Stubblefield, one year before the accepted father of radio, Nikola Tesla (1856-1943), unveiled his own invention. Stubblefield's subsequent fate is not a happy one; despite producing and selling a telephone system, he went bankrupt, retreated into a hermit-like existence and eventually starved to death.

Nikola Tesla is an intriguing character who holds the US patent for the invention of the radio. However, this accolade was restored to him posthumously, and the holder of the patent from 1904 to 1943 was the Italian inventor Guglielmo Marconi (1874-1937). Tesla invented the Tesla coil, which is still used in radios and televisions today, and got his radio an American patent in 1897, though this was subsequently overturned and given instead to Marconi in 1904. Tesla's genius as an inventor is undisputed, and he made many contributions to science, but his eccentricities in the end worked against him. He claimed that he had picked up extraterrestrial signals while stationed in Colorado Springs, which he believed originated from Mars, and also to have successfully designed and tested a 'death ray' which could be used to erect electromagnetic fields capable of destroying aircraft and ships. He also died impoverished, having sold his patents and seriously compromised his reputation, though his work was celebrated again shortly after his death.

Guglielmo Marconi for a time was thought to be the real father of the radio. Born in Italy in 1874, he patented the first radio in Britain in 1896 and established a radio station on the Isle of Wight. He said that 'radio was born to the sound of a rifle shot', as this was how his assistants signalled they had received a transmission, and in later years he became seriously involved in Italian fascism – in fact Mussolini was his best man. In 1904 he was awarded the US patent over Tesla, possibly due to his American backers, who included Thomas Edison. He also proved that radio waves could be transmitted not only in a straight line, but could 'bounce', though the experiment in which he did this has recently been contested; some now believe that Marconi picked up background radiation and mistook it for a signal in Morse code. Most famously, Marconi transmitted a message from the US president to the king of England in 1903. He died in 1937 (upon his death radio stations around the world observed a two-minute silence) so did not have to suffer the indignity of seeing his patent returned to Tesla, who is now widely regarded as the most influential contributor to the invention of radio.

The other most notable inventor to have furthered wireless technology was Reginald Fessenden (1866-1932), who believed radio could work better as continuous waves rather than the sparking method employed by Marconi. By 1903 he had designed and built a high-frequency alternator, and on Christmas Eve, 1906, he transmitted the first audio radio broadcast (on which he read from the Bible and played the violin) from Brant Rock, Massachusetts, to ships at sea.

RAYON

The origin of the world's first manufactured fibre can be traced back to the speculations of the Englishman Robert Hooke (1635-1703) in 1665. As curator of the New Royal Society of England, Hooke observed through his examination of silk that it was possible to spin a similar fibre from glutinous substances. His musings were over 200 years ahead of the commercial production of rayon, a fabric first described to the public as artificial silk.

In the meantime, others continued along the natural route. In 1709 the French naturalist, François Xavier Bon de Saint Hilaire (1678-1761), sought to create a fibre from spiders' web that would match the luxurious quality of silk. He began by boiling and washing cocoons and then drew threads from them using fine-toothed combs. Though the process was particularly

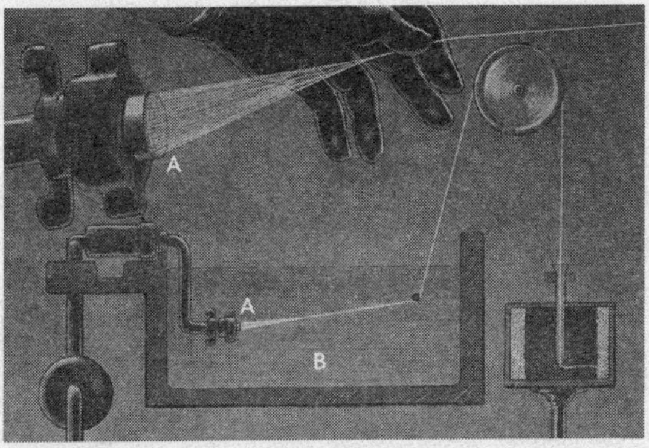

Wet Spinning

This method of manufacturing Rayon is still used today. Liquid cellulose passes through the tiny openings of the spinneret (A, shown in detail at the top) to produce fine filaments that solidify after contact with an acidic chemical bath (B).

time-consuming, François was able to produce socks, gloves and even stockings, the latter weighing only two and a quarter ounces. Despite the overpopulation of spiders in the world, this method was doomed impractical. It was estimated that one kilogramme of silk required 1.3 million spider cocoons.

In 1764, French scientist René Antoine Ferchault de Réaumur (1683-1757) suggested in his book *Histoire des Insectes* that silk is 'nothing more than dried-up liquid gum' and could surely be manufactured by combining it with resin. His method for creating a silk fibre involved forcing varnish through tiny holes in the bottom of tin cans, which produced filaments that hardened in the air. Though considered to be as impractical as the use of spiders' web, this basic mechanism would be mimicked by researchers in the future.

The true pioneer of rayon was the French chemist Hilaire de Charbonnet, Comte de Chardonnay. He was a former pupil of the great bio-chemist Louis Pasteur (1822-1895), whose contribution to science included research into the diseases of the silkworm. Charbonnet was more than likely inspired by the work of his mentor. In 1864 he began studying the metamorphosis of the silkworm's only form of sustenance, the mulberry leaf, into the raw silk produced by its salivary glands. Charbonnet performed experiments in attempt to replicate this process, but his efforts failed. Yet he knew that the main ingredient in the mulberry leaf was cellulose. In 1884 he hit upon a landmark combination: he dissolved nitrocellulose (cellulose exposed to nitric acid) into alcohol and ether. The result was a new fibre, which he described as 'an artificial textile material resembling silk'. Charbonnet refused to call his discovery 'artificial silk', since what he had actually created was something entirely new.

Nonetheless, the fibre that was invented in 1894 by DuPont scientists Charles Cross (1855-1935), Edward Bevan and Clayton Beadle was marketed as 'artificial silk'. While Charbonnet's fibre had serious limitations – for starters, it was highly flammable – these men were able to produce a filament using a method that was quite similar to the earlier process invented by Réaumur. The more technologically advanced procedure known as 'wet spinning' begins with purified cellulose. It is converted to liquid form and then passed through a 'spinneret', which functions something like a showerhead. The cellulose is fed through its tiny openings to create fine filaments, which solidify after direct contact with an acidic chemical bath. The versatile fabric that we know as 'rayon' continues to be manufactured in this way today.

RAZORS

Shaving has been a gentleman's duty – and a lady's chore – since the dawn of civilised society. Egyptians were obsessed with cleanliness and it was customary for men and women to strip their bodies entirely of hair. The women shaved their heads completely, using razors with gold and copper blades, before replacing their natural hair with elaborate, alluring wigs. Other early materials used to rid the body of hair were sharpened flint, sharks' teeth and shells – the latter imitating modern-day tweezers by clinching the hair and plucking it out from the root.

In 1770, French artisan Jean-Jacques Perret published a complete shaving guide called *La Pogonotomie*. Shortly afterwards, he produced the Perret razor. Effectively a facial switchblade, the razor could be folded into its own wooden handle. Though it met with some success, the device was cumbersome and dangerous to say the least. Beyond the foldaway design, the L-shaped instrument was sorely lacking in anything we'd consider to be safety features. The central hinge could loosen, leaving the blade to flop about in nastily unpredictable ways, and fumbling with the razor while still half-asleep in the morning was a hazardous exercise.

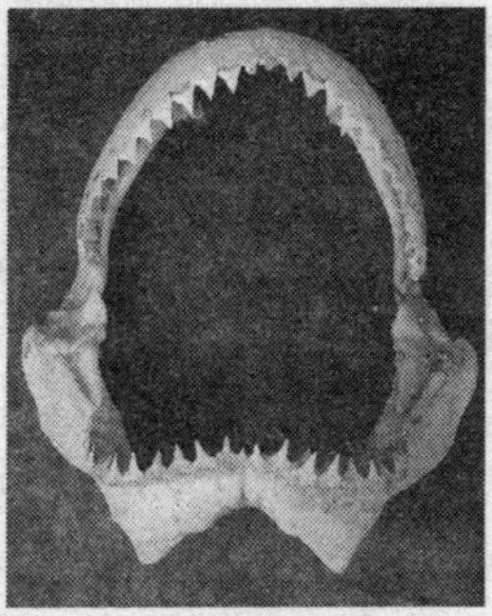

Fancy trimming your goatee with these? Sharks' teeth were among the earliest materials used to shave hair.

William Henson improved the situation in 1847 when he placed the blade perpendicular to the handle, transforming its usability and reducing the need to hold it quite so gingerly. Shaving became a more entertaining prospect, and barbers reaped the rewards. However, this didn't lead to a full-scale curb of the Perret 'straight razor'; some men still use a straight razor today, as they feel it achieves a closer shave than any commercial product.

The real breakthrough came at the dawn of the twentieth century. The superbly named King Camp Gillette (1855-1932) from Wisconsin was inspired by the success of the throwaway Crown Cork bottle cap and, in 1901 – with the help of Massachusetts Institute of Technology graduate William Nickerson to develop a lightweight blade – he perfected the first cheap, disposable razor. Production began in 1903, and his invention took the shaving business by storm. Gillette razorblades were issued en masse to the British army during World War I and you will be pleased to learn that frontline troops were respectably shaven before all major battles. Baron

Marcel Bich (1914-1994) – world-renowned inventor of the Bic biro – followed Gillette's lead in 1975, fusing the blade and grip into a fully disposable unit. This was a step up from Gillette's long-term handle for short-term blades (and a good deal safer, too). Bich further revolutionised the product by making the handle from thin plastic, making it cheap to manufacture and to buy.

REFRIGERATOR

For some people it is hard to imagine life without the fridge freezer. To be able to take a dish of food from frozen to (microwave) cooked in minutes, this would have seemed the stuff of crazy science to our nineteenth century ancestors.

The process of refrigeration is based on the theory that absorption of heat by a fluid (refrigerant) as it changes from a liquid to a gas lowers the temperature of the objects around it.

The first man to discover this process was William Cullen (1710-1790) at the University of Glasgow in 1748, but he did not use the technology to develop anything. Then in 1805, American inventor Oliver Evans (1755-1819) invented the first refrigeration machine. The first patent for mechanical refrigeration was issued to another American, Jacob Perkins, in 1834.

One of the first industries to realise the benefit of refrigeration was the brewery industry. Refrigeration enabled breweries to sell a uniformly cold product all year round. By the 1870s nearly every brewery in the US was equipped with refrigerating machines. The meat-packing industry also quickly realised the advantages of refrigeration.

In 1911 General Electric launched their first home-market style refrigerator, but most machines were powered by motors so large they had to be housed in a separate room, so were not exactly convenient. The main problem the industry faced was the coolants, which on occasion leaked and proved fatally poisonous. It was not until the 1930s when the use of chlorofluorocarbons was developed, that refrigerators became more suitable for the consumer market. Refrigerators as a domestic item began

ORIGINAL FACTS

Musical fridge anyone?

At the time of writing, a new fridge called the thermoacoustic chiller is about to be launched in America in conjunction with Ben & Jerry's ice cream. The technology uses sound waves to power it, doing away with damaging emissions. It does not even damage your eardrums, as it emits decibels high above the human range of hearing. The principle on which it is based is that sound waves can change the temperature of whatever they travel through. Potentially this could be a more much more green way to power a whole household once the technology behind it has been improved.

to take off, and a whole refrigerated cuisine evolved as manufacturers produced menus demonstrating all that a refrigerator could do. Prior to refrigerators, frozen desserts and salads had been available only to the rich, but now this lifestyle was available to the middle classes. What could be more high class than frozen cheese salad, or an icy frappé made of condensed tomato soup?

By the mid-1950s, over 80% of Americans owned refrigerators. However, the new chlorofluorocarbon technology may have been less damaging to humans, but it was having a catastrophic effect on the environment. Not until the 1970s, did people realise that chlorofluorocarbons were literally eating the ozone layer. Today, they are banned in the production of all refrigerators.

RING-PULL

Prior to the 1960s, the only way to open a canned drink was to punch a hole in the top. In fact, you needed to punch two holes, one to let the liquid out and another to vent the can to allow it to flow. One day in 1959, Ermal Fraze of Dayton, Ohio found himself without a suitable implement and attempted to open a can with his car bumper. Most of his drink turned to foam and was lost. Fraze, who owned a local tool manufacturing business, meditated on the problem. He decided that the solution was to attach an opening lever to the lid of a can. He eventually came up with a tab that ruptured a pre-scored tear strip when pulled. The aperture that resulted was large enough to allow air to enter the can at the same time as the liquid poured out. Unfortunately, Fraze's tab was difficult to manipulate. In 1965, Omar Brown and Don Peters improved the design by replacing it with a ring that also functioned as a lever. Ten years later Brown joined forces with Fraze to create the 'push-in' ring-pull. This eliminated the problem of sharp-edged litter and remains the mechanism of choice.

ROLLER SKATES

In 1760, a Belgian instrument maker named Jean-Joseph Merlin (1735-1803) attended a masquerade party thrown by a Mrs Corneily at Carlisle House in London. On his feet he wore a novel pair of boots equipped with small metal wheels. Merlin made a grand entrance, skating across the ballroom floor while playing the violin, but then he rather spoiled the effect. As one contemporary account had it, 'not having provided the means of retarding his velocity or commanding his direction, he impelled himself against a mirror of more than 500 pounds value, dashed it to atoms, broke his instrument to pieces and wounded himself severely'.

ROULETTE

The great French mathematician Blaise Pascal (1623-1662) is said to have invented a primitive version of the roulette wheel in the 1650s during his search for a perpetual motion machine. Another intriguing theory holds that the game was devised by the Chinese and brought to Europe by Dominican monks. What is certain is that a version of roulette was already a popular carnival entertainment by 1842, when the Frenchmen François

Gambling is a series business: here a casino employee tests the roulette wheel with a spirit level to ensure absolute accuracy before the day's gambling begins.

and Louis Blanc introduced the element that transformed roulette into the definitive casino game. By adding a '0' to the wheel, they provided the all-important house edge. But the 2.7% advantage this bestowed was not enough for American casino owners. When roulette arrived in the US (in time for the 1849 Gold Rush) they added a '00' slot, thereby doubling the house margin to 5.3%. For this reason, the 37-hole European version of the game has tended to appeal more to serious gamblers than the 38-hole American equivalent.

François Blanc built some of the first casinos in Monte Carlo and grew rich on the profits. The legend that he sold his soul to the devil was probably started by a disgruntled gambler who had noticed that the numbers on a roulette wheel added up to 666.

RUBBER BAND

In 1823, the British inventor Thomas Hancock made the first crude rubber bands from a machine he had constructed called the Masticator. As vulcanised rubber had yet to be invented, they were inclined to become brittle as a result of cold temperatures. Nevertheless, Hancock formed a lucrative partnership with Charles Macintosh (1766-1843), the creator of the waterproof garment of the same name. Elastic bands made from vulcanised rubber, designed to hold papers and envelopes together, were patented by Stephen Perry of the London manufacturers Messrs Perry & Co on 17 March 1845.

RUBIK'S CUBE

During the late 1970s, all nerds and quite a few 'ordinary' people were to be found fiddling with multi-coloured cubes. The device that so monopolised their attention was invented by a Hungarian interior design lecturer named Erno Rubik (b.1944). He had the perfect pedigree for an inventor: his mother was a poet and his father was an engineer. In 1974, Rubik became obsessed with the technical problem of building a block of cubes which could be manipulated to change the position of any of its elements. He tried a system based on rubber bands and then realised that the solution lay in the shape of the non-surface faces of the individual cubes. By moulding them with complex patterns of grooves and projections, he was able to build a fully mobile cube of the kind he had been looking for, with three rows of three smaller cubes on each of its faces. Rubik painted each face a different colour and started to twist. Then he spent a month trying to get the cubes back to their original positions.

Rubik applied for a Hungarian patent for the cube in 1975 and received one two years later. Initial sales were modest. Then an entrepreneur called Dr Laczi Tibor saw a waiter playing with one of the puzzles and tracked down the inventor. He persuaded both Rubik and the Hungarian authorities to let him take the cube to the Nuremberg Toy Fair in Germany. When he got there, his tactic was simply to wander around the stalls twiddling away at his product. It worked: a British toy salesman named Tom Kremer placed an order for a million.

Rugby school, 1906, with its pupils playing the sport of the same name outside.

RUGBY

The game of rugby is named after Rugby school, a boys' school in Warwickshire, England. In the 1800s, seven schools, including Rugby, were playing variations on football and rules were introduced to formalise the game into 'association football', which has since become football.

However, Rugby school was playing the game with a noticeable difference – running with the ball. The exact story of how the game evolved is not known for definite, but the popularly held belief is that in 1823, a Rugby pupil named William Webb Ellis (1806-1872), in a moment of disregard for the rules, picked up the ball and ran with it.

The existence of hand ball games can be traced back to many ancient tribes, such as the Celts, Maoris, Faroe Islanders, Philippines and Eskimos. The Celts played the game of 'Caid' (Cad), with the interesting, rather colourful meaning of a bull's scrotum. It has been suggested that Webb Ellis was influenced by his father, who would have played Cad in the Dragoons Guard in Ireland. Tempting as this story may be, there is little evidence to support it. Webb Ellis was certainly a student at Rugby between 1816 and

1825, but his act of picking up the ball was never recorded by anyone. The first mention of it was in the Rugby school magazine in 1886, by an old Rugbeian, M Bloxham. However, the story really only amounts to hearsay, as Bloxham was not there and no one else can corroborate the account.

What we know for certain is that by the 1830s, the main characteristics of today's rugby were being played at Rugby school. Running with the ball, scrummaging, and high goal posts with a cross-bar 10 feet above the ground had all been introduced. In 1871 the Rugby Football Union was formed by 21 clubs mainly based in southern England. By 1875, rules were introduced if no goals were scored. From 1876 three tries equalled one goal in points and this gradually moved to giving more value to the scoring of tries, so that by the mid-1890s, the scoring was much closer to what we know today. Despite its southern roots, it became increasingly popular in northern England throughout the 1890s.

Today, two forms of the game exist; Rugby League and Rugby Union. Rugby Union has 15 players, while Rugby League has only 13 players. In Rugby League there are two substitutions permitted and play stops after every tackle, whereas in Union, play is continuous. For the more faint-hearted there is touch rugby, a less strenuous and physical game, with less likelihood of the players emerging with misshapen ears.

S

SAFETY PIN

The safety pin was invented by a mechanic from New York named Walter Hunt (1796-1859). Prior to the creation of the device that has secured a billion nappies, Hunt had devised a flax-spinning machine, a repeating rifle, a nail-making machine and several other innovative items. Unfortunately, he never managed to make any money. The closest he had come was the invention of America's first sewing machine in 1834, but he had declined to patent or pursue the idea for fear of putting seamstresses out of work. Hunt's safety pin was another case of missed financial opportunity. He came up with the idea in 1849 while agonising about a US$15 debt to a friend. Hunt sat down and resolved to invent something lucrative before he got up again. He began toying with a piece of brass wire and soon realised that he had a simple fastening device in his hands. Three hours

US Patent 6,281 10 April 1849

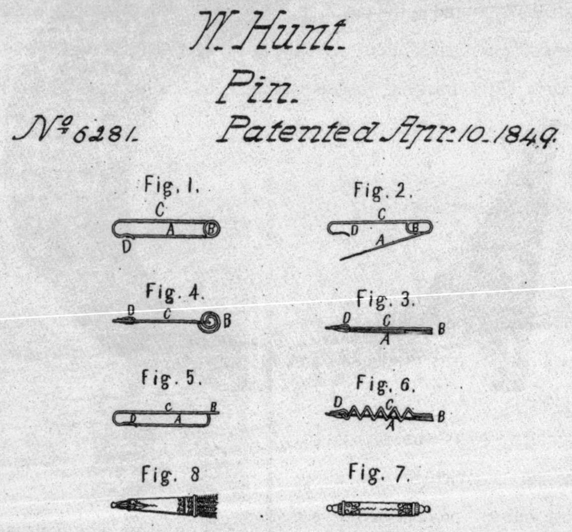

The original patent, with a number of different designs, as well as the familiar one we know and love.

later, he had written a patent application. It described the item in the following terms: 'The distinguishing features of this invention consist in the construction of a pin made of one piece of wire or metal combining a spring and clasp or catch, in which catch, the point of said pin is forced and by its own spring securely retained.' Hunt promptly sold the patent application and the model safety pin he had made to support it for US$400 to the friend he owed the US$15 to. He missed out on millions but was able to repay his US$15 debt.

SARI

Universally associated with the Indian subcontinent, the sari (or saree) is an adjustable garment made in varying lengths that can fit women of all sizes. However, it is not merely a piece of fabric, but part of a tradition that stretches back some 5,000 years.

The first recorded evidence of the use of saris dates from 3,000BC: statues discovered in the region that was once ancient Sumer – home to the Indus valley civilisation – have the flowing coverings carved into them. Ancient Greek travelling records give nods to beautiful fabrics worn by Indian women. Later, the sacred 1,200BC Sanskrit text, *Rig Veda*, contains references to celestial weavers making a bright, golden material that is turned into a sari.

The sari plays a prominent part in many aspects of Indian life, including the arts.

It is also in the *Rig Veda* that the devotional aspects of the sari are explored. The verse entitled 'Samhita' treats the process of weaving, of which the sari is the most notable product, as an act with religious purpose: 'Sacrifice resembles a loom with threads extended this way and that, composed of innumerable rituals. Behold now the fathers weaving the fabric; seated on the outstretched loom. "Lengthwise! Crosswise!" they cry.'

In addition to its appearance in the *Rig Veda*, the sari also plays a prominent part in the seminal Indian mythic poem, 'Mahabarata'. Detailing an epic struggle between the Pandava and Kaurava families, the poem features a scene in which the Kauravas attempt to humiliate Drapudi of the Pandavas by divesting her of her sari – but, thanks to an intervening Lord Krishna, the robe just grows longer and longer as it is pulled from her body.

With themes of religion and myth so prevalent around the sari, it is likely that it evolved from a simple and convenient mode of dress to a means of conveying and perpetuating religious beliefs and mythic traditions with the aid of a wide variety of patterns and designs. Today it still has a symbolic function: to communicate a woman's social class or marital status according to the style, colour and quality of the cloth.

SCRABBLE®

Scrabble is a game that combines chance, knowledge, inspiration and a relentless trial-and-error perseverance, all of which is similar to the story of how it was invented. The mind behind Scrabble was that of Alfred Mosher Butts (1899-1993), an unemployed architect who turned his active, inquiring brain to the task of designing a popular game in the early 1930s. 'I didn't have anything to do,' said Butts in 1986, 'so I thought I'd invent a game.'

Butts deduced that all games fall into three broad categories: games of chance usually involving numbers, such as dice and cards, board games such as chess and draughts, and word games such as crosswords or anagrams. Word games were hugely popular at the time, so he began with that, but eventually integrated chance and numbers into his game, which he spent years designing and modifying. His first effort was called 'Lexiko'; there was no board, but Butts had already cracked perhaps one of the most important aspects of Scrabble – the frequency of the letters. He calculated this by scanning newspapers and journals and working out the best

US Patent 2,752,158 26 June 1956

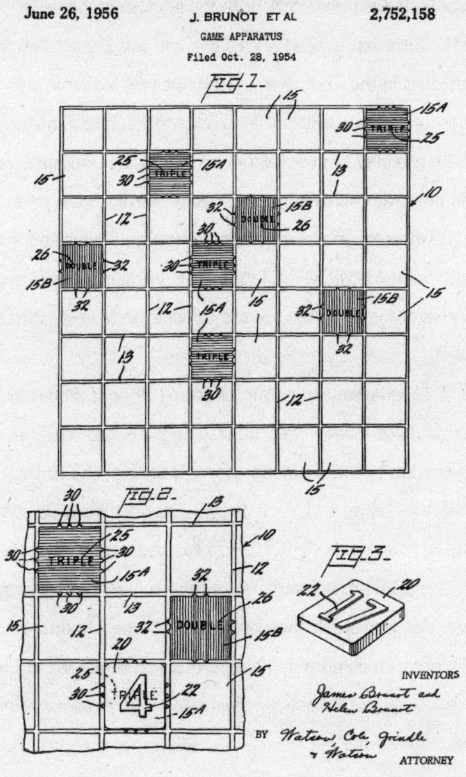

The scrabble game board patent, filed in James Brunot's name.

quantity of letters needed to construct the widest number of words. He made a number of sets himself, but his design was rejected by the Parker Brothers and other toy manufacturers of the day.

However, Butts was not easily dissuaded. Sensing that a board would provide a greater focus to the game, he created one, and gave each letter a score based on its regularity (or rarity). He also dropped the hand of letters from nine, as used in Lexiko, to seven, and began experimenting with double and triple word scores. He called his new game 'Criss-

Crosswords', and although it enjoyed some popularity in the northeast, he again received rejection letters from the games companies.

By 1947 Butts was back at his old job, and was less interested in getting the game established. That year, a lawyer named James Brunot came on to the scene, who was interested in buying the rights. Butts agreed on the grounds he received a royalty on every copy sold. Under Brunot's control, some important changes to the game were made; the rules were simplified, the arrangement of the 'premium' squares was altered, the 50-point rule for using all seven letters was introduced and, perhaps most importantly, the name was changed to Scrabble because Brunot liked the sound of it.

At first sales were slow, but then in 1952, the owner of Macy's department store – apparently a Scrabble fan – was outraged to discover the game wasn't stocked in his store. Macy's placed a large order and promoted the game, and soon Brunot couldn't keep up with demand, enlisting Selchow and Righter Co (who had previously rejected Butts' design) to make and distribute the game. In 1953, a million games were sold, and the next year that figure rose to nearly four million. Today Scrabble is reportedly found in one in every three American homes. Alfred Butts was never made rich by the game, but remains beloved among hardcore Scrabble players. He died in 1993.

SCUBA DIVING

The word Scuba stands for 'self-contained underwater breathing apparatus'. The key to the 'self-contained' part is the demand regulator, a mechanism that ensures that air is delivered to divers' lungs at the same pressure as the water surrounding them. This is important as water pressure doubles between the surface and a depth of 10 metres, increasing by one atmosphere every 10 metres thereafter. If divers were to breathe air at surface pressure while at depth, the contents of their lungs would simply buoy them upwards. This would not only defeat the object of the exercise; it would make them vulnerable to the dreaded 'bends' (a

US Patent 1,370,316 1 March 1921

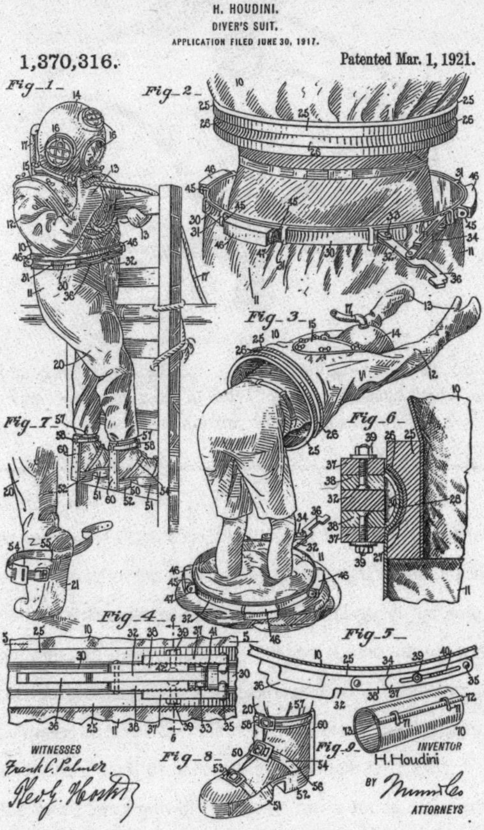

Harry Houdini even has a patent for a diving suit. The point of his design, appropriately enough, was to allow the diver to escape the suit should their air supply fail.

condition in which nitrogen bubbles form in the body as a result of over-rapid ascent).

Prior to World War II, the pressure problem placed severe restrictions on the viability of diving. The sport had been practiced in various forms since

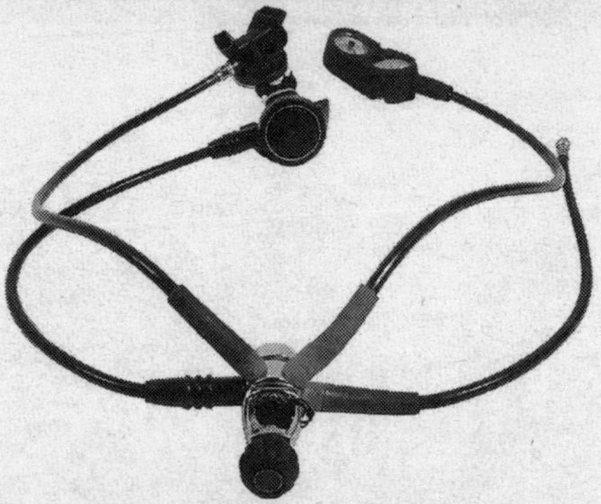

Scuba diving equipment, including the demand regulator which ensures air is delivered to the diver at the same pressure as the surrounding water.

1788, when the British engineer John Smeaton (1724-1794) refined the diving bell by inventing a pump system that delivered air to the diver via a hose pipe, but it was limited by the need to maintain a connection with the surface. Autonomous systems were developed but they were never entirely satisfactory. One example was an apparatus devised by the Englishman William James in 1825. It allowed divers to breath via belts of weighted tanks of compressed air attached to their copper helmets, but only gave them about seven minutes of diving time. Divers were finally liberated when the Frenchmen Emile Gagnan (b1900) and Jacques Cousteau (1910-1997) invented the demand regulator in 1942-3. Their method involved redesigning a pressure regulator that Gagnan had originally made to allow cars to run on vegetable oil.

SEAT BELT

The invention of the seat belt predated it becoming a standard automobile fixture by more than 80 years. The first American patent for a seatbelt was issued on 10 February 1885 to Edward J Claghorn of New York City. The

US Patent 312,085 10 February 1885

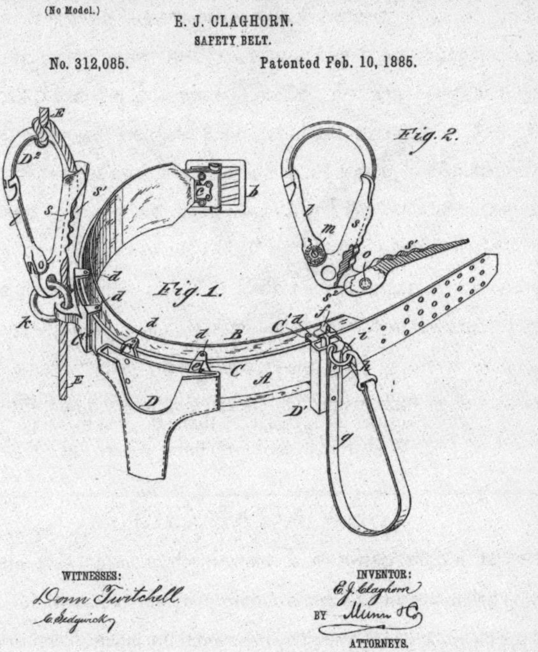

The first American seatbelt patent, bearing more resemblence to a climbing harness than today's modern devices.

patent application described the device as 'designed to be applied to the person, and provided with hooks and other attachments for securing the person to a fixed object'. Seat belts were first used in quantity during World War I, when they were installed in military aircraft. It took the automobile industry some time to catch on. In 1956, Ford equipped some of its cars with the quick-release AutoCrat safety belt patented by the brothers Kenneth and Bob Ligon, but the first manufacturer to include them as standard was Volvo in 1959, opting for Swedish inventor Nils Bohlin's (1920-2002) modern, three-point design. The presence of seat belts in cars has been required by US law since 1968.

SELF-TANNING LOTION

During the 1920s scientists discovered an interesting property of a sugar cane derivative called dihydroxyacetone (DHA). When rubbed onto human skin it triggered a protein reaction that imparted a brown colour. In the late 1950s, a mail order 'doctor' named John Andre discovered the effect in an old medical journal. He had been looking for a self-tanning product for years. His New York firm Drug Research, Inc. promptly developed a DHA-based formula called Man-Tan. As the name suggests, Man-Tan, which came to market at the start of 1960, was targeted firmly at males. The product was far from perfect – unless skilfully applied it left users with conspicuous brown collar marks and blotchy orange skin – but the prospect of an effortless tan was too much for many Americans to resist. Six million four-ounce bottles were sold in the first six months alone.

SEWING MACHINE

The first recorded patent for a sewing machine was in 1755. The British patent was issued to a German named Charles Weisenthal, for a needle that was designed for a machine. However, the patent failed to describe any other information regarding the machine itself.

It was not until the end of the eighteenth century that another attempt was made, this time by Thomas Saint, an English cabinetmaker. He was issued the first patent for a complete machine in 1790, although his patent drawings failed when they were reproduced.

Numerous attempts were made over the next 50 years to produce a fully functioning sewing machine, but these all proved unsuccessful.

It was not until 1830, that the French tailor Barthelemy Thimonniere (1793-1857) invented a successful machine which replicated the chain stitch used for embroidery by using only one thread and a hooked needle. Unlike the others that had gone before him, he was able to convince authorities of the usefulness of the machine. In less than 10 years he had a factory running, with around 80 of its machines dedicated to stitching uniforms for the French Army. However, his invention was not welcomed

An old-style sewing machine, powered by a foot treadle and using a needle that moves up and down (rather than side to side, as was previously the case).

by Parisian tailors and he was almost killed when a group of them burnt down his factory.

Meanwhile in America, the sewing machine was catching on. Elias Howe (1819-1867) is often attributed with the invention of the sewing machine, as he was the first to utilise the eye pointed needle and shuttle system. He patented his machine in 1846, but had trouble marketing it and protecting his invention from others who were developing innovations of their own.

Isaac Singer (1811-1875) built the first commercially successful machine, which used a needle that moved up and down instead of side to side and was also powered by a foot treadle (previous machines had been hand-cranked). He founded a company, which became the world's largest manufacturer of sewing machines by 1860.

However, Singer's machine used the same lockstitch that Howe had created and Howe sued Singer for 'patent infringement' in 1854. He won and gained his rights to a share in the profits of this invention. In 1863, Howe made up to US$4,000 a day from his rights. In about 1857, Willcox and Gibbs began manufacturing chainstitch sewing machines, which gradually became more popular than the Singer machines.

The sewing machine was a success, despite early fears about its effect on womenfolk who would use it. At first it was thought, in the mid-nineteenth century, that women would be too excitable to manage such an instrument. Then there were the worries as to the kinds of activity women would indulge in if they weren't making bedding and knitwear, with card-playing and shopping of primary concern. It was once suggested to Singer that, 'You want to do away with the only thing that keeps women quiet – their sewing!'

SHAMPOO

The word shampoo first entered the English language around 1762, although the product, of course, is much older. The word is of Anglo-Indian descent, from the Hindi word *champo* meaning 'to press, knead the muscles, or massage'. In Eastern cultures, such as India, various natural ingredients had been used to clean the hair for centuries. Indonesian shampoo was traditionally made from rice husk and rice straw. These were burnt to form and ash, which was mixed together with hot water to form a lather. Coconut oil was then added to the hair to put moisture back into it.

In Western cultures, there was at first not much difference between soap and shampoo. English hairdressers used to boil soap in soda water and added herbs to give hair health and fragrance. The first commercially successful shampoo was developed by John Breck in the US in 1908. His advertisements featuring the 'Breck Girl' became very popular during the 1930s, and many famous Hollywood starlets posed for his ads. This was also the era when the much gentler PH-balanced shampoos entered the market, spearheaded by Breck, but increased competition in the 1970s diminished the brand's status.

SHOPPING MALL

The title of the world's oldest shopping mall is hotly disputed and ultimately depends on definition. The strongest candidate is probably the Stoa of Attalos, a two-storey colonnaded building built in the Athens market place in the second century BC. It contained 42 rooms, all of which were used as shops and businesses. The multi-storey Trajan's Market in the Forum in Rome is almost as old, having been built by the architect Appollodorus of Damascus between 100 and 112AD. Nineteenth century claimants to the title include the Galeries Royales St-Hubert built in Brussels in 1847 and Milan's cross shaped and glass roofed Galleria Vittorio Emanuele, built between 1865 and 1877. The issue is easier to resolve if we bring automobiles into the equation. The first purpose-built shopping mall with parking spaces to appear in America was the Country Club Plaza outside Kansas City, Missouri. It was built by the JC Nichols Company and opened in 1922.

SHOPPING TROLLEY

The shopping trolley, or shopping cart as it is known in the US, was invented in 1937 by Sylvan Goldman (1898-1984), the owner of a chain of supermarkets in Oklahoma City. Goldman was moved to action by the sight of customers struggling with overloaded baskets. With the help of a mechanic named Fred Young, he knocked up a prototype cart from a folding chair, a pair of wheels and a couple of wire shopping baskets. At first Goldman's innovation was resisted by men, who thought pushing a trolley effeminate, and women, who considered it unstylish, but the saving of energy was too much to resist. By 1940 there was a seven-year waiting list of other stores clamouring for the carts.

SILK

Most people agree that silk and silk production (or sericulture), has its origins in ancient China. Legend has it that it was discovered by Yuen Fei, the wife of the Yellow Emperor who ruled in China about 2,600BC. The story goes that she dropped a cocoon in some hot tea and saw a single strand unwind into a thread. A group of woven threads and fragments have been found in the Zhejiang province dating back to this time and evidence of sericulture has been found even earlier in sites along the lower Yangzi river.

There are numerous different types of silk moth found in different countries, but the Chinese species it originated from is the blind, flightless moth; the Bombyx Mori. It lays 500 or more eggs in five to six days and dies soon after. From one ounce of eggs about 30,000 worms are produced, which eat a ton of mulberry leaves and produce about 12 pounds of silk in their lifetime.

Silk production is a lengthy and laborious process, which in part explains its expense. Conditions need to be just right to prevent the moth from hatching. The eggs must be kept at 65°F and in ancient China, Chinese peasants used to keep the eggs next to their bodies, or in blankets under their beds to keep them warm. Once the eggs have hatched, the worms feed day and night on handpicked mulberry leaves until they are fattened up and have enough energy to store up for the cocoon stage. The cocoons are made from a jelly type substance secreted from their silk glands, which hardens when it comes into contact with the air. They spend three or four days spinning a cocoon around themselves until they look like fluffy white balls.

Once the cocoons have matured in a warm, dry place for eight or nine days, the cocoons are ready. They are steamed, or baked to kill the worms and then dipped in hot water to loosen the tightly woven filaments. Each cocoon is made of a filament between 600 and 900 metres long. Five or eight of these super fine filaments are twisted together to make one thread. The process of sericulture was a closely guarded secret in China, and anyone caught trying to smuggle silkworm eggs or cocoons out of the country was put to death.

The famous global silk route began in the second century BC and stretched 4,000 miles through South Asia and the Middle East. No one knew that it began in China until the thirteenth century when Marco Polo (1254-1324) made the journey along the whole route into the interior of China.

The three stages of the silk moth lifecycle:

1. Silk worms gorging on mulberry leaves.

2. The fluffy white cocoon, from which silk is made.

3. One of the numerous different types of silk moth.

The secret was bound to get out at some point, and around 200BC the Koreans began producing silk, when Chinese immigrants arrived there. India began making its own in 300AD, and by the sixth century the Persians had also mastered the art of silk weaving. After Marco Polo had opened up the route to the interior of China, silk production was established in Italy with the aid of 2,000 skilled weavers transported from Constantinople. Today world silk production has doubled in the last 30 years, despite the development of man-made fibres, and China continues to be the dominant producer.

SILLY PUTTY®

The world has the Japanese to thank for the invention of Silly Putty. When Japan invaded the rubber-producing nations of South East Asia during World War II, the US faced a sudden shortage of this essential commodity. The War Production Board responded by issuing a plea to the nation's industries to come up with a synthetic alternative to rubber. One day in 1943, a Scottish engineer named James Wright, who was working on the problem in General Electric's laboratory in New Haven, Connecticut, mixed some silicon oil and boric acid in a test tube. The result was a strange gooey substance technically described as 'visco elastic'. Wright started playing around with the material. When he threw a lump on the floor, he was amazed to find that it leapt straight back up at him.

For seven years, the strange substance circulated among chemists, but no one could think of a practical use for it. In 1949, the product attracted the attention of Ruth Fallgetter, the owner of a New Haven toy store, who called in a marketing consultant named Peter Hodgson (1912-1976) to help her sell it. Although the bouncing putty was the second most popular item in Fallgetter's annual catalogue, she lost interest, leaving Hodgson to pick up the baton. In 1950, he thought up the name Silly Putty and had the idea of selling it in little plastic eggs. That August, *New Yorker* magazine ran a story on the novelty item. Within three days, Hodgson had received orders for a quarter of a million units.

SKATEBOARDING

The story goes that skateboarding grew out of the 1950s surfing culture, when surfing enthusiasts began nailing the bases of their roller skates to the front and back of wooden planks. Unstable though they were, these crude constructions allowed them to 'sidewalk surf' down the street and downhill. The craze soon caught on and in major cities around America 'street surfing' became the new cool, with enthusiasts introducing challenges such as curb jumping and vaulting banked surfaces.

By the early 1970s, bicycle manufacturers and toy companies were producing stable unbreakable boards, to meet the growing demand. As technology improved, so did skater skills, and a standard set of manouevres evolved such as the 180- and 360-degree turns. Skaters initially used empty construction sites and public spaces to practise their skating skills, but gradually purpose-built skating parks began to emerge where formal contests could be held. Contests are now held internationally and the international skateboarding circuit is a serious competitive sport.

After a lapse in popularity at the end of the 1970s, skateboarding has recently made a comeback as both a pastime and a sport, thanks in part to the popularity of its fellow sport of snowboarding. Between 1997 and

2000, skateboarding was ranked as America's sixth largest participant sport, and is likely to remain a central part of American youth culture.

SLICED BREAD

Whenever a popular new product is launched, it is hailed as 'the best thing since sliced bread'. Given the proverbial excellence of the pre-cut loaf, it is surprising that the man who invented it does not enjoy a higher profile. His name was Otto Frederick Rohwedder (1880-1960) and he lived in

US Patent 1,867,377 12 July 1932

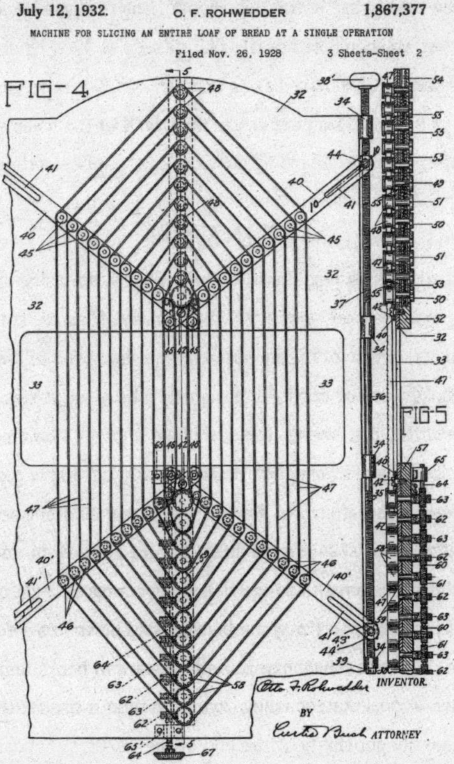

Otto Rohwedder's bread-slicing machine.

Davenport, Iowa. Rohwedder patented a bread-slicing machine in 1912 but it took 17 years for him to find a commercial taker. During this period he was beset by personal and technical problems. In 1915 he was diagnosed with a fatal kidney disease and given a year to live. He was too ill to work for seven years and in 1917 his newly-built factory in Illinois burned down, destroying all his prototypes. When Rohwedder recovered, the concept proved a hard-sell. The bakeries he contacted were not impressed by the fact that his original machine held cut loaves together with metal hat pins. He also had a terrible time persuading them that pre-sliced bread would sell, as they were understandably concerned that sliced loaves would go stale faster than intact ones. Rohwedder's solution was to develop a machine that wrapped bread as fast as it sliced it. In 1929 he obtained his first customer, the Chillicothe Baking Company in Chillicothe, Missouri. The following year he sold his patents. By 1933, 80% of the bread sold in the US was pre-sliced.

SLINKY®

A famous advertising jingle asks, 'What walks down stairs, alone or in pairs, and makes a slinkity sound?' The answer, of course, is the Slinky. More than 300 million of the animated springs have been sold worldwide, but the Slinky was not originally designed to be a toy. It was meant to stabilise nautical instruments. The story begins in 1943 on a boat in the Cramp Shipyard in Philadelphia. While working on a spring system to counterbalance the effects of waves at sea, a marine engineer named Richard James (1914-1975) accidentally knocked one of the components off a shelf. He was amazed to see the spring taking on a life of its own. Instead of just falling to the ground, it marched down in a series of arcs. It alighted on a lower shelf, moved on to a stack of books, stepped onto the desktop and climbed down to the floor. Then it neatly reassembled itself into an upright coil.

When James brought the spring home to show to his wife Betty (b1918), she recalled thinking, 'Oh boy, here we go again.' She was well used to his outré contraptions, which included a compressor that pumped soda from the basement into the family fridge. But after Richard had spent a year tinkering with the design she became convinced of the moving coil's potential as a toy. To find an appropriate name for the device, she settled down with a dictionary and selected the adjective 'slinky', Swedish for 'sinuous'. The Slinky was unveiled at Gimbels Department Store in Philadelphia in 1945. The Jameses sold 400 in an hour and a half. In 1960, Richard abandoned his family and the Slinky empire for a religious cult in Bolivia, but Betty stayed on as CEO, later setting up a Slinky factory in Hollidaysburg, Pennsylvania.

THE SMILEY EMOTICON

The grinning yellow symbol that is now used in emails to denote jokes or amusement has an interesting history. It was a hippie icon, appearing on more than 50 million badges sold by 1971, and was later adopted by the rave generation. Drug dealers even used it to adorn ecstasy tablets. But the origins of the Smiley are thoroughly establishment. During the 1960s, the bosses of the State Mutual Life Assurance Company of Worcester, Massachusetts, identified a problem. The company had recently merged with the Guarantee Mutual Company and staff morale had plunged. As a consequence, they needed something to cheer up their employees. They commissioned a graphic designer named Harvey Ball (1921-2001) to devise a cheerful symbol for use on company stationery. Ball came up with the Smiley, for which he was paid $240. He never earned another penny from his invention but remained true to the philosophy it embodied. 'Hey,' he used to say, 'I can only eat one steak at a time.'

SLOT MACHINE/ONE-ARMED BANDIT

The forerunner of the one-armed bandit was the 'drop card' poker machine that emerged on the East Coast of America during the 1890s and later became popular in San Francisco. It consisted of an almost full deck of cards adhered to five revolving drums. The 10 of spades and jack of hearts were usually removed by the owners to reduce the odds of getting a Royal Flush. Although the drums were set spinning by the introduction of a nickel, the machine was not a true one-armed bandit as the astronomical number of possible permutations prohibited automatic payout. Instead, winners had to show their 'hands' to the proprietors of the establishment in question. They were then given free drinks or cigars according to the house system of rewards.

The fully fledged one-armed bandit was developed around 1895 by Charles Fey, a Bavarian-born mechanic who lived in San Francisco. It had three reels, each adorned with spade, heart and diamond symbols plus one cracked Liberty Bell, from which it took its name. As a result of this arrangement, Fey's invention had far fewer possible combinations than the 'drop card' machine. This made automatic payout possible. The jackpot hand of three bells in a row triggered a princely reward of 50¢. The original Liberty Bell machine is on display at the Liberty Belle Saloon & Restaurant in Reno, Nevada.

SMS/TEXT MESSAGES

SMS technology was built into the first generation of mobile phones almost as an afterthought. Although the concept was first proposed in 1982 by a Finnish civil servant named Matti Makkonen (b1952), it was developed during the late 1980s by a group of Norwegian engineers working on a European telecommunications project called Global System for Mobile Communications (GSM). The idea was to equip the phones with a simple message system that worked even when they were switched off or out of range. The first SMS message is thought to have been sent on 3 December 1992 by Sema Group engineer Neil Papworth. Using a PC, he texted his Vodafone colleagues' mobile phones to wish them a 'Merry Christmas'.

SOAP

According to legend, soap takes its name from a mountain called Sapo where the Ancient Romans once sacrificed animals to the gods. (*Sapo* is the Latin word for soap). Rain water supposedly carried wood ash and fat from the offerings down to the River Tiber, where they mingled with clay deposits. In time, the local women are said to have discovered that washing clothes with the resulting mixture produced sparkling results. The problem with the theory is that there is no historical record of any peak called Mount Sapo.

The oldest evidence suggests that soap was invented by the Babylonians. A soap-like substance has been found in excavated clay cylinders dating from 2,800BC. Inscriptions on their exteriors reveal that the material was made by boiling animal fat with ashes. They do not, however, reveal how the soap was used. The first reference to soap in a European language was made by Pliny the Elder (7BC-53AD) in the first century AD. He described Germanic tribes using it to lighten their hair.

SOCKS

The word socks is derived from the Latin word *soccus*, which evolved into the Old English *socc* and the Middle English word *socke*. Roman comic actors wore *socci* as shoes. Romans also sometimes wore them with the sandals in which they conquered much of the known world, which could be where the much-feared socks-and-sandals fashion faux pas first arose.

By the fifth century AD, sock-like garments called *puttees* were worn by holy people in Europe to symbolise purity. By 1000AD, they had become an indication of wealth among the nobility. During the Crusades and Gothic periods, linen hose for men was made from leg-shaped pieces sewn together and tied at the waistband.

From around 1340 onwards, gentlemen of the upper class began to wear shorter body garments called 'gippons', or 'pourpoints', which exposed their nether regions. Strangely it was entirely proper for the aristocracy to parade themselves in this way if they so desired, but anyone else would have been fined. The tailors of the day not surprisingly began to focus on developing higher quality legwear to cover the gap between hose and tunic, if only to keep the daft toffs warm.

An advert for garters. Men used to wear these small elastic straps to keep their socks from falling down (much as women wear suspenders to keep their stockings up), though the practice has fallen out of favour in recent times.

By the fifteenth century, 'stockings' described a covering for the feet or legs, and were usually knee-length. As coverings for the more temperature-sensitive male areas crept back into fashion, giving us britches and trousers, stockings shortened as if to welcome the new sartorial sense: they'd done their bit.

In 1589, the Reverend William Lee (1550-1610) of Nottingham invented the first knitting machine. This meant that socks were easier to produce and increased their popularity, so they were no longer simply a luxury item. Socks enjoyed immense success in the seventeenth century, when embroidered socks became highly fashionable.

The development of nylon in the 1930s revolutionised the fabrics industry, including the lowly sock. Previously socks had been made only out of silk or cotton, but now they could be made from mixed fibres more cheaply. Even more significant, was the effect nylon had on the stockings and tights industry (see Nylon). Today, socks come in all shapes, sizes and colours depending on your mood; and, thanks to modern textile technology, many have antibacterial capabilities. Just don't, whatever you do, wear them with your sandals.

SOUP

Soup was so popular with our prehistoric ancestors that they made it before they invented vessels sufficiently robust to cook it in. The standard

A wartime advert suggesting Heinz soup as a solution to rationing.

method appears to have been to heat stones in fires and then drop them into clay bowls filled with liquid food. The word 'soup' is derived from the 'sops' of bread that served as plates during in the Middle Ages. They were later placed at the base of bowls to soak up the juices of whatever food was served in them. Eventually the sops disappeared but the name was retained to describe meals with a liquid consistency.

SPAM

SPAM® originated as a luncheon meat brand introduced by Hormel Foods in 1937. The name represented the first and last two letters of the main ingredients, namely 'spiced meat and ham'. The delicacy was later eulogised in a tongue-in-cheek song by the Monty Python team. For obscure reasons, in the early days of the internet someone decided that 'spam' would make an appropriate name for unsolicited email.

The first email spam is believed to have sent by a Digital Equipment Corporation marketing rep on 1 May 1978. The message invited every member of ARPANET living on the West Coast of America to attend one of a series of receptions at which the company's new DEC-20 machine was to be unveiled. The poster was chastised for using the ARPANET directory linappropriately. On 18 January 1994, every newsgroup on another forerunner of the Internet called USENET received a message reading: 'Global Alert for All: Jesus is Coming Soon'.

SPANDEX

Spandex, the essential stagewear of eighties rock musicians, was spawned by the American DuPont company in 1959, who were also responsible for the invention of CFCs. Their brand new fibre went on to revolutionise clothing in the post-war era.

Spandex – or elastine rubber, as it is known in other parts of the world – is made from prepolymers, colorants and stabilisers. Research and development on the textile started during World War II, in the search for a replacement for rubber. At the time, rubber was being used in the majority of war equipment, making its price wildly unstable. Originally codenamed 'Fiber K' by DuPont inventor Joseph C Shivers, spandex could be stretched up to 600 times its length and still recover its original shape. Once it went into production it was given the more delicate-sounding brand name, Lycra®.

Since the 1950s, it has been made into swimming costumes, wetsuits, leggings and all manner of sportswear, thanks to its lightweight composition and its imperviousness to perspiration, detergents and body oils. Spandex even attracted its own fetishists, thanks to the material's skin-tight attributes and its shiny aesthetic, reminiscent of leather or PVC.

SPEED DATING

In 1998, a Los Angeles Rabbi named Yaacov Deyo devised a new way for Jewish singles to meet other Jewish singles with a view to marriage. Together with his wife Sue and his students at the Aish HaTorah Jewish resource group, Deyo developed a round robin system which allowed every male who turned up to a meeting to have a brief 'date' with every female and vice-versa. The concept rested on two convictions: first impressions count and comparison helps people learn how they feel about each other.

The first Speed Dating event was held at Pete's Café in Beverly Hills at the end of 1998. The system was refined the following year to consist of seven seven-minute interviews. Although Aish HaTorah owns the service-mark for Speed Dating, the concept has been adopted all over the world, proving particularly popular in urban centres and among the gay community.

SPELLING BEE

Two questions come to mind regarding the literacy contests known as spelling bees. First, why on earth are they named after honey-making insects? Second, how and when did they come into existence? The short answer to the first question is that they aren't. The word bee in this sense is derived from the English dialect word *been*, which is itself a variation on *boon*, meaning voluntary help given to farmers by their neighbours at harvest time. In time both the final letter and the connotation of assistance were dropped. By the middle of the nineteenth century people were using bee to refer to any communal gathering with a focused purpose. There are contemporary accounts of knitting bees, corn husking bees and even lynching bees.

Turning to the second question, the birth of the spelling bee depended on the introduction of standardised spelling. In the US, the reference book responsible was Noah Webster's *The Elementary Spelling Book*, of which 80 million copies were sold between 1783 and 1883. The first references to spelling bees appear in American literature during the 1840s, although at this stage they were known as 'spelling matches'. The term 'spelling bee' was first used during the 1870s, the decade in which the craze really took off. In April 1875, the *Oakland Daily Evening Tribune* claimed that 'The spelling fever is playing bob with our pet phrases; "too diaphanously attenuated" is now the substitute for "too thin".' The American National Spelling Bee was started by *The Louisville Courier-Journal* in 1925. The first winner, Frank Neuhauser of Kentucky, triumphed by correctly spelling 'gladiolus.' The current sponsors EW Scripps assumed the role in 1941. Of the 81 champions to date, 43 have been girls and 38 boys.

SPIN DRYER

A Frenchman named Pochon invented the forerunner of the spindryer in the early 1800s, a hand-cranked clothes dryer that treated fabric with all the delicacy of a rotisserie. The clothes were wrung manually and put into a simplistic metal drum with ventilator holes that one then rotated over a fire. Unfortunately, although they came out dry, the clothes also came out smelling of smoke, covered in soot and sometimes burnt.

The mangle was invented around 1861 – a manual wringer that squeezed out excess water by crushing the clothes between two sandwiched rollers. The clothes then had to be air dried to remove the rest of the moisture. Mangles were later attached to washing machines ('machines' being a loose term for buckets with foot-pedals that swirled clothes around...) but proved somewhat perilous; the first powered ones had no safety mechanism, and if a finger/sleeve/strand of hair was caught in the mangle there was no stopping it.

Joseph Huebsch of Wisconsin tried to solve the problem by inventing the first open-ended steam dryer in 1931. Huebsch and his eight brothers started a commercial laundry in 1891 and the Huebsch Manufacturing Company is still around today, manufacturing machines for all kinds of laundry needs and washing the clothes of America. But the key story is that of J Ross Moore. Some say that he wanted to keep his mother from

An early washer-dryer! This has a washing drum at the bottom and a mangle at the top, through which clothes were wrung by hand. The two sandwich rollers squeezed out excess moisture, and the clothes were then air-dried to remove the rest.

having to dry the family's clothes in the freezing North Dakota winter; others claim that he was saving himself from his own hated daily chore. Either way, he spent 10 years developing a dryer that started out as a rack in front of a fire and ended in 1936 as an electrical spinning drum with hot air blasting through it. After trying for many years to have his idea endorsed by manufacturers who said it wouldn't sell, he finally had an offer from Hamilton's Manufacturing Co, who helped him develop and build his invention. The drying machine was finally released to the public in 1938. Moore travelled America on a Hamilton's promo tour of his new appliance, and now over 78% of American households have one. Modern additions to his original model include electric buttons and thermostats, or humidity sensors that detect when the clothes are dry and shut the machine down automatically.

Tumble dryers use the most amount of energy in the home out of all household appliances. Spin dryers, however, use centrifugal force, as opposed to evaporation, to remove water from clothes and are so energy efficient that they make tumble dryers blush with embarrassment – they remove the same volume of liquid in two minutes as a tumble dryer does in 20, with G forces of around 1,420. It is claimed that a spin dryer pays for itself in under a year in electricity savings, and it is rare to find a washing machine without a built-in spin cycle.

SPRAY GUN

The spray gun was developed not to paint cars or for the benefit of graffiti artists but to squirt medication into hard to reach nasal passages. It was invented in 1888 by Allen deVilbiss of Toledo, Ohio. He had concocted a special medicated oil for the treatment of nose and throat complaints and needed a way to deliver it. Two years later he founded the DeVilbiss Company to manufacture medical atomisers. His son Thomas, who became a partner in the firm in 1905, was responsible for developing new applications for the technology such as the DeVilbiss perfume atomiser which was introduced in 1907. During the 1920s, the company began to supply spray guns to the furniture and automobile industries. The devices dramatically reduced the drying times needed for paint and lacquer.

SPRING MATTRESS

By modern standards, beds were pretty uncomfortable prior to 1871. In that year, a German named Heinrich Westphal invented the innerspring mattress. Hitherto, beds had typically been stuffed with horsehair, wool, down or straw. Westphal's design made novel use of the steel coil springs developed during the industrial revolution. They had been patented for use in the seats of armchairs as early as 1857, but no-one had thought of using them in mattresses. Despite his invaluable contribution to world restfulness, Westphal died in poverty, having omitted to patent his invention.

STAPLER

The world's first stapling machine was made for the exclusive use of King Louis XV of France (1710-1774), who reigned from 1715 to 1774.

Documents had previously been fastened with ribbons passed through slits in their upper left hand corners, or by a combination of ribbon and sealing wax. Louis's staples were handmade and inscribed with the royal insignia. According to some accounts, they were made of gold and encrusted with jewels.

Modern staplers emerged towards the end of the nineteenth century. The first device to insert and clinch staples simultaneously was patented by Henry R Heyl of Philadelphia on 25 September 1877. A machine which held multiple staples in a cartridge was introduced the following year.

US Patent 195,603 25 September 1877

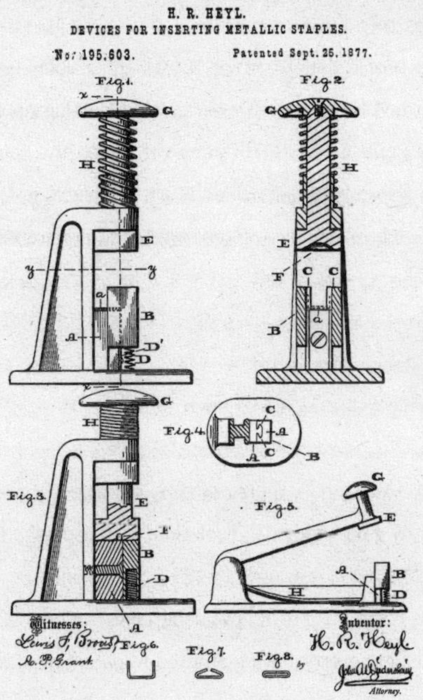

Henry Heyl patented the first stapler to insert and clinch staples simultaneously, as modern staplers do.

STETSON

For understandable reasons, the invention of the 10-gallon hat or Stetson is usually credited to the man who lent it his name. John Batterson Stetson (1830-1906), the son of a hat maker, was born in 1830 in Orange, New Jersey. When his father died, he continued in the family business working for his brothers. He planned to set up on his own but was stricken with tuberculosis and moved to St Joseph, Missouri for health reasons. In 1862, Stetson accepted an invitation to join a gold prospecting team heading for Pike's Peak. One night, the men were soaked by a storm and the untanned animal skins they had lashed together to improvise a tent were ruined. This prompted Stetson to show his colleagues how to make felt, a trick he had picked up in the hat business. He shaved the fur from some hides, beat it until the hairs had separated, then soaked and boiled it. Then he shaped the resulting mass into a wide brimmed waterproof hat. He wore it for the remainder of the trip. When he reached Pike's Peak, Stetson sold the hat to a cowboy for $5. After failing to strike gold, he decided to go back to his former trade. He established a business in Philadelphia and made a batch of felt hats based on the one he had made on the mining expedition. Then he headed west to market them. He sold his first 'Boss of the Plains' hat in Century City, Colorado in 1865. In time, they became known as Stetsons.

What the standard story fails to mention is that Stetson fought a long and unsuccessful battle for the patent rights to the hat. He was ultimately forced to pay licensing fees to a firm based outside Bristol in England called Christy's. The company had been making wide-brimmed felt hats for sugar workers in the West Indies since 1812.

SUDOKU

You either love or hate it, but either way you probably assume that the numbers game Sudoku was devised in Japan. The name, which roughly translates as 'number only once', reinforces this belief. But in fact Sudoku was invented by a retired American architect named Howard Garns, and first appeared in *Dell Puzzles and Word Games* magazine in 1979. Garns, who published the game anonymously under the name 'Number Place', drew his inspiration from the Swiss mathematician Leonhard Euler (1707-1783), who devised a similar game called 'Latin Squares'. Garns' innovations were to split Euler's grid into nine squares and add the rule that any number can only be used once in each square.

The game first appeared in Japan in April 1984 in a magazine called *Monthly Nikolist*. It is probable that it was in that country that the game first became enormously popular for linguistic reasons. As Japanese characters are symbolic rather than phonetic, they are not suitable for use in western-style crosswords. Everyone, however, needs something to do while sitting on a commuter train.

SUGAR

Sugarcane is generally believed to have been domesticated in New Guinea some 8,000 years ago. By 500BC, the plant had travelled to India, where the locals had learned to produce crude crystallised sugar by evaporating the sap. They called the substance *sakar*, meaning 'grain'. Alexander the Great's general Nearchos, who led an expeditionary force into the sub-continent around 325BC, described sugarcane as 'a reed that yields honey without bees'. The next great sugar pioneers were the Arabs, who established plantations on several Mediterranean islands by 1000AD. Crete was home to a particularly important refinery, as the Arabic name for the island suggests. They called it Qandi, which is the origin of the term 'candy'.

Australian workers posing with sugarcane, the source of the sap from which sugar is made.

Sugar arrived in Western Europe in the early Middle Ages in the luggage of returning crusaders. It was considered a miraculously healthy substance and was extremely expensive. At first, only monarchs could afford to be conspicuous consumers. As late as the fifteenth century, a teaspoon of sugar in Britain was worth £3 in modern terms. It was only after sugarcane found a fertile home in the Caribbean that the price fell within the reach of ordinary people.

THE SUIT

The evolution of the men's suit into the modern two- or three-piece garment we know today is usually attributed to English gentlemen and London tailoring fashion in the late eighteenth century. However, its gradual emergence dates further back to the origins of tailoring, when the cut and shape of clothes became important. The Italian Renaissance saw the start of a more fashion conscious era. The loose robe was shortened and stitched to fit the contours of the body, although it caused a stir amongst the Church, who saw tight-fitting clothes as indecent.

France and Spain entered into the quest for the new haute couture, but England was slow to catch on. It wasn't until Charles II (1630-1685) returned from his exile in France that the template for the modern suit was born. Influenced by the court of Louis XIV (1638-1715) at Versailles, he decreed in 1666 that men at court should wear a long coat or waistcoat, a cravat (precursor to the modern necktie), a wig and breeches or trousers gathered at the knee.

A gentleman in the court of Louis XIV. The fashions he saw here inspired Charles II to demand a similar style of dress in English courts.

It was the Americans who introduced the final changes to tailoring to give us the modern day suit. Looking for an alternative to the long heavy frock coats worn for formal attire, men started to wear lighter coats cut just below the waist when not on business. This has now developed into the 'lounge suit' in Great Britain, or the 'business suit' in North America.

The tuxedo was created as an informal suit for evening wear at the end of the nineteenth century. As it grew in popularity, it gradually became acceptable formal wear and has practically replaced the white tie, which is now seen only at the most formal occasions.

SUNGLASSES

The Inuit have been wearing anti-glare goggles for centuries and perhaps millennia to counter snow-blindness. The Canadian Museum of Civilisation in Quebec owns an example that dates from about 1200AD. It is made of walrus ivory and features narrow slits designed to restrict the amount of light reaching the pupils.

The earliest spectacles with tinted lenses were designed not to protect the wearer from the sun but to help Chinese judges maintain an inscrutable demeanour. Their facial expressions were concealed by means of lenses made from smoked quartz. The devices were in widespread use by the twelfth century. The first tinted glass spectacles were made around 1752 by an English scientific instrument maker named James Ayscough, but like their Chinese predecessors they were not designed with sunbathing in mind. Ayscough was purely interested in the correction of vision. He believed that light passing through clear glass was harmful to the eyes and advocated the use of green or blue lenses instead.

In the modern era, the wearing of sunglasses was pioneered by the Hollywood actors of the early twentieth century. Once again, however, the sun had little to do with it, neither were the stars in question seeking anonymity or attempting to hide cocaine-dilated pupils. Instead, they were seeking respite from studio lights, which were excessively bright in the early days of the film industry. 'Shades' only assumed their contemporary function when sunbathing came into fashion during the 1920s and 1930s. Mass-produced sunglasses were introduced to America by Sam Foster under the brand name Foster Grant. They were first sold in 1929 in an Atlantic City branch of Woolworths.

SUNTAN LOTION

For much of history, the suntan was stigmatised due to its association with manual labour. Things changed, however, when improved transport and growing affluence led more and more people to vacation in exotic locations. Suddenly, tanned skin became associated with wealth and cosmopolitanism. The trend was greatly enhanced during the 1920s when the fashion setter and designer Coco Chanel (1883-1971) inadvertently acquired a deep tan while on holiday in the French Riviera. The other major development was the introduction of skimpy swimsuits during the 1930s. Victorian and Edwardian bathers had covered themselves up so thoroughly that they couldn't have tanned even if they had wanted to.

Although the new fashion created a huge market for products that would protect the skin while allowing sunbathers to tan in safety, the earliest effective sunscreen was developed by the military. In 1944, a pharmacist called Benjamin Greene developed a petroleum-based product called 'red vet pet' (red veterinary petrolatum) to protect soldiers operating in desert regions. It was thick and sticky but it worked, as Greene demonstrated by applying the lotion to his own bald patch.

SUPER GLUE™

Super Glue is based on a substance called cyanoacrylate, which was synthesised in 1942 by Harry Coover (b1919), a scientist at Eastman Kodak. He was looking for suitable material for clear plastic gun sights for the war effort, but the product was unworkable as it stuck to everything it touched. During the 1950s, Coover and his aptly named colleague Dr Fred Joyner developed cyanoacrylate into a workable and immensely strong glue. It was patented in 1956 and first sold two years later as 'Eastman 910 adhesive'. The product immediately attracted media attention, notably in

1958, when Coover appeared on the television show *I've Got a Secret* and arranged for host Gary Moore to be hoisted off the ground via a single drop of the miracle glue.

Since it arrived on the scene, the qualities of Super Glue have made it the subject of numerous urban myths. Some of the tallest tales turn out to be true. Combatants in the Vietnam War really were issued with tubes of the glue to use in the event of serious flesh wounds, and a Peruvian man starved to death after sealing his jaws by accidentally getting some of the stuff on his teeth. Handle with care...

SUPERMARKET

The supermarket depends on the concept of self-service. Prior to World War I, grocery stores invariably operated on a 'full service' basis. Customers

US Patent 1,242,872 — 9 October 1917

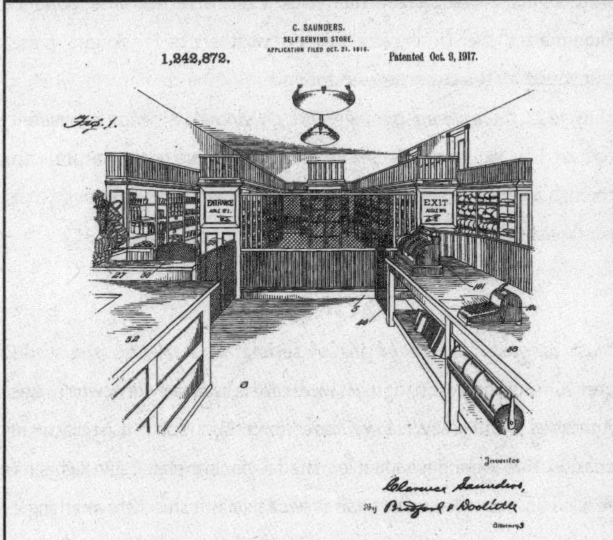

Clarence Saunders' patent for the first self-service grocery store, the wonderfully-named 'Piggly Wiggly' at 79 Jefferson Street in Memphis, Tennessee.

would present their lists of desired items to a clerk stationed behind a counter, who would then fetch and pack them on their behalf. Then, in 1916, a man named Clarence Saunders (1881-1953) had a brainwave. Why not get customers to do the fetching and carrying for themselves? On 6 September, Saunders opened the world's first self-service grocery store at 79 Jefferson Street in Memphis, Tennessee. It bore the splendid name 'Piggly Wiggly'. When Saunders was asked to explain why, he simply replied, 'so people will ask that very question'.

Customers at the store entered through a turnstile and selected goods from shelves as they made their way round a maze of narrow aisles. Then they progressed to an unheard-of phenomenon called a 'checkout stand'. The new style of shopping had profound implications for the advertising business. Whereas grocers had previously tended to sell only a couple of brands of any given product, that selection was now in the hands of the shopper, and product recognition took a quantum leap in importance. Supermarket shelves became packed with brightly coloured brands competing for the customer's attention.

By 1923, there were 1,268 Piggly Wiggly stores in America, the majority operated by franchisees. Saunders contrived to lose control of the chain through an ill-timed attempt to buy up all the company's shares, but he continued to work on new retail concepts up to his death in 1953.

SURFING

Most people think the origins of surfing lie in Hawaii, the world's premium surfing destination. However, there is also evidence to suggest it originated in Peru. What we know for certain is that it began within cultures that were dependant on the sea for survival. Centuries ago in Peru, in a place called Chan Chan, it became a ritual for fisherman to ride the waves to shore on their one-man reed canoes after fishing. This may have been purely for fun, or a way of increasing their knowledge of the sea and water skills. Either way, it is a tradition that has endured and is still practised today.

In Hawaii, the natives rode huge surfboards they called Olos. These early surfboards were carved out of the wood from native Koa trees. It was a time-consuming process to make these boards, as each one had to be hand-carved using primitive tools. The surface was then smoothed out by rubbing pebbles over them and finally they were treated with natural oils to protect the wood and keep the water out. They were often in excess of 12 feet and weighed up to 70 pounds. The longer boards were usually reserved for the royalty, the alii, and the shorter boards were used by the commoners, or makaaaina.

Excerpt from a promotional leaflet issued by the Hawaii Promotion Committee.

However, when the missionaries arrived in Hawaii, surfing was banned as it was thought to be immoral. Surfing did not regain its popularity until the early 1900s, when Duke Kahanamoku (1890-1968), known as the father of surfing, took the world by storm with his amazing surfing talent.

Born in Hawaii, Duke Paoa Kahanamoku (nicknamed 'Duke' after the Duke of Edinburgh's visit to the island), surfed with his brothers and entertained tourists with their tandem rides. Initially it was his swimming prowess that brought Duke fame and recognition; he broke numerous world records and won an Olympic gold medal in 1912. As he toured the world with his swimming contests, he also took the opportunity to display his surfing talents. In 1914, he wowed onlookers at Freshwater Beach in Australia, when he rode the waves effortlessly with his board made from a local sugar pine. Gradually the surfing bug caught on and Hawaii became a surfing hotspot. In 1965, The Duke

Kahanamoku Invitational became an international event, drawing in aspiring surfers.

Today, thanks to the development of high-density foam, surfboards are quick to make and come in a variety of styles and shapes, which can be tailored for the more advanced or more novice.

SWIMSUIT

The history of the swimsuit is an interesting insight into social prudishness through the ages. If we go back to our Greek ancestors, it seems they were remarkably ahead of the times. Early artwork found in Italy and Greece reveals that some women wore a two-piece outfit reminiscent of today's bikini.

The concept of public bathing did not take off in modern times until the eighteenth century, when it became popular to take to the waters in spa towns such as Bath, England. However, this was a far more prudish affair than the liberal Greeks and propriety over comfort was the order of the day. Woollen swimming costumes were the norm, as they did not become see-through when wet. Instead, they simply clung to the body and felt horrible.

Not only were skimpy swimsuits unacceptable in the nineteenth century, women also used bathing machines to travel to the water's edge, so preserving their modesty.

Women wore shirt-type dresses and stockings, even weighing the skirt ends down to stop them lifting up in the water, which can only have been a safety hazard.

By the nineteenth century, seaside holidays were becoming more common as train travel to the US and UK coasts improved. Bathing machines – mobile changing rooms for the beach – were invented, which carried women to the water's edge to preserve their modesty.

When Australian swimmer Annette Kellerman wore a one piece costume like this in 1907, it proved so scandalous she was arrested, though the subsequent display of public support for her indicated a growing change in society's prudish views on swimwear

In 1907, the history of the swimsuit took a leap forward when Australian swimmer Annette Kellerman (1887-1975) wore a one-piece costume that revealed her arms and legs to a swimming demonstration in Boston. It proved so scandalous that she was arrested. The groundswell of public support for Kellerman, and the general indignation over her indecency charge, was the beginning of the end of prudish views on swimwear. Over the next 20 years, the one-piece design formed the basis for the standard costume, and the development of synthetic fibres such as Lycra® helped to create a more functional and well-fitting costume.

With the feminist revolution of the 1920s, styles became more risqué and revealing, with lower necklines and shorter skirts. The pace of design calmed down during the war years, as people were less inclined to go for public swims. After the war however, liberalism triumphed and redesigning the swimsuit was taken up with renewed vigour. In 1946 the ultimate in minimalist swimwear appeared (see Bikini).

SYNTHESIZER

This versatile instrument is largely responsible for the rise of electronic music, and made a significant contribution to the advancement of many other genres. The most famous synthesizer is the Moog, invented by Robert Moog (1934-2005) – pronounced like 'vogue' – who pioneered groundbreaking circuit technology that allowed him to blend, control and modulate various sound wave forms. 'Synths' became available in the mid-1960s (The Beatles and Mick Jagger bought expensive, early models), and over the years have been redesigned and vastly improved into today's more or less exclusively digital versions, operating as interfaces for computer programs.

Some have traced the origins of the synthesizer much further back, where its evolution becomes intertwined with that of the organ and keyboard. The timeline stretches as far back as the third century BC, when a Greek engineer designed a primitive version of the pipe organ using a system of pumps, air chambers, water and pipes. Another notable example was William Duddell's (1872-1917) 'Singing Arc', the first electronic instrument. In 1899 Duddell, a physicist, was asked to investigate why carbon arc lamps made a variety of sounds during use, ranging from a hum to a high-pitched whistle. He discovered that the more electricity applied to the lamp, the higher the sound emission. By connecting a keyboard to the light system at the London Institute of Electrical Engineers, he performed a remarkable demonstration in which all the lamps in the Institute were made to sing. His discovery was not followed up until Moog's work over half a century later.

In 1963 Robert Moog was completing a physics PhD at Cornell University and selling Theremin kits in his spare time. A Theremin is an unusual instrument consisting of two aerials, controlling pitch and volume, that are operated by moving one's hands within the proximity of the aerials. Originally intended for use in classical music, the Theremin was

Kraftwerk were just one of the electronic music pioneers who could not have existed without the synthesizer.

famously adopted by Brian Wilson (b1942) of The Beach Boys. Walter Sear, Moog's New York representative for Theremins (who would later gain infamy as a pornographer), introduced him to Herb Deutsch, and together they experimented with the circuits that would become the first synthesizer. They promoted their designs at conventions and received a few orders. However, things did not really get moving until Moog met the transsexual musician Wendy (formerly Walter) Carlos (b1939), who had a big influence on the designs: 'She really understood instinctively what I was doing right and wrong. The fixed filter banks came from Wendy. Lots of other things, too; I've lost track,' said Moog. Carlos recorded the first full-length Moog album 'Switched-On Bach', which created enormous publicity for the synthesizer.

By 1970, other companies had begun to manufacture synthesizers, and throughout that decade the popularity of the instrument grew, contributing to the avant-garde 'prog-rock' movement, as well as instilling the first lessons of minimalist electronic music through the innovations of groups such as Kraftwerk.

In retrospect, the synthesizer can be said to have had social effects too. It was used by much of the Black Futurist movement to give their music a cosmic, alien atmosphere, which reflects their experience of social isolation; indeed Lee Perry (b1936) said of his recording studio: 'It was like a space craft. You could hear space in the tracks.' This work fed directly into the emergence of the Detroit techno scene in the early 1980s, which used synths to create a dystopian, urban sound, and inspired artists across Europe and the US to create their own brands of techno, house and dance music. These in turn fathered the most recent avant-garde electronica practiced by the likes of Aphex Twin. Having lived to see his invention revolutionise the world of music, Moog died of a brain tumour in 2005.

T

TABLE FORK

Although the Romans sometimes used forks as serving implements, it appears that forks were first used by individual diners in Byzantium in the Eastern Mediterranean. The Cleveland Museum of Art owns what may be the earliest example, a six inch Byzantine silver fork dating from the fourth century.

When table forks appeared in Europe, they were considered the height of pretentiousness. It is likely that the implement was brought to the continent in the eleventh century by a Turkish princess who had married Domenico Selvo, the heir of a Venetian Doge (the city's ruler) in 1070. Instead of eating with her fingers, the princess insisted on using a little two-pronged golden fork. Her behaviour scandalised the people of Venice, particularly the clergy. They protested that God had provided humans with natural forks in the form of their fingers, and that to use artificial ones was therefore to insult him. When she died, the Bishop of Ostia was moved to write 'of the wife of the Venetian Doge, whose body, after her excessive delicacy, entirely rotted away'.

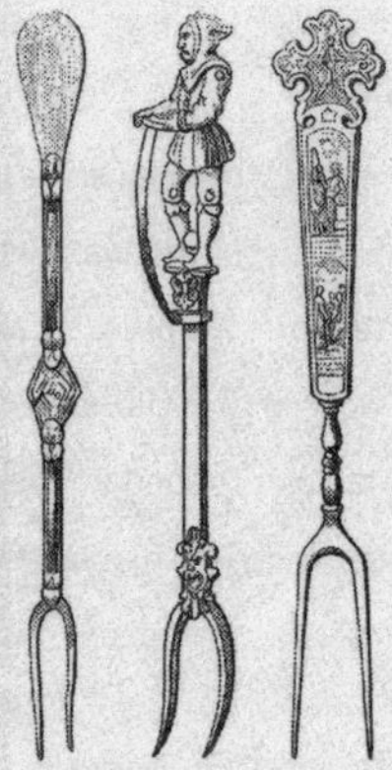

A collection of old, elaborately-decorated forks.

TABLE TENNIS/PING PONG

Table tennis appears to have been created by British Army Officers in 1881. They erected a 'net' across a table using open books, carved a ball from a Champagne cork and used cigar box covers as bats. Upper-middle class Victorians were soon using their dining tables as mini-tennis courts for an after-dinner diversion. In the early days, table tennis was given a number of onomatopoeic names, such as 'gossamer', 'whif whaf' and 'flim flam'. In 1901, the name 'Ping Pong' was registered as a trademark by the English games manufacturer J Jaques & Son Ltd, and it caught on, perhaps as a result of its pleasingly oriental sound. The company later sold the trademark to Parker Brothers in the US. The other important development of 1901 was the introduction of celluloid balls, which an English table tennis enthusiast named James Gibb had discovered on a trip to America. Two years later, EC Goode produced the first modern bats, which he made by sticking sheets of pimpled rubber to light wooden 'blades'.

TABLOID NEWSPAPER

The concept of the tabloid newspaper was thought up by the pioneering British journalist Alfred Harmsworth (1865-1921) (later Lord Northcliffe). The term 'tabloid' was borrowed from the pharmaceutical industry. The format was first used for a one-off edition of the *New York World* published on the first day of the twentieth century. It was edited by Harmsworth on the invitation of the paper's owner Joseph Pulitzer, who had been impressed by his work in the UK. In 1903, Harmsworth launched the world's first regular tabloid in London. *The Daily Mirror* was about half the size of a standard broadsheet newspaper and was initially aimed at women.

The first American paper published in tabloid format was *The Illustrated News*, founded by Joseph Medill Patterson (1879-1946) in New York in 1919. The paper provided readers with a lurid diet of sex, crime and scandal. After a few months its name was changed to the *New York Daily News*.

TATTOOS

The art of decorating skin with permanent pigment goes back to the Neolithic era. In 1991, a 5,300-year-old mummified corpse was discovered in the Tyrolean Alps between Austria and Italy and christened 'Otzi'. Fifty nine separate tattoos were found on the body of the 35-year-old man. Most consisted of parallel lines but some were in the shape of crosses. In 1891, archaeologists discovered relatively complex tattoos on the skin of an Egyptian priestess who lived some time between 2,160 and 1,994BC. Inscriptions on the mummy revealed that her name was Amunet.

TEDDY BEAR

A commonplace household name, who would have thought the teddy bear had a presidential past? The famous namesake was none other than President Theodore Roosevelt (1858-1919). Strangely, stories surrounding the origin of the Teddy Bear coincide on both sides of the Atlantic. The US origin lies in a cartoon by Clifford Berryman, which appeared in the *Washington Post* on 16 November 1902. Titled 'Drawing the Line in Mississippi', the cartoon depicted Roosevelt refusing to shoot a baby bear. Roosevelt had been visiting Mississippi to try and resolve a border dispute it was having with Louisiana. A hunting trip was organised in Roosevelt's honour. However, the hunting was so poor that someone captured a bear and tied it to a tree for Roosevelt to shoot. Upon seeing the bear, Roosevelt refused to shoot it, claiming it was unsportsmanlike. Berryman's cartoon satirising the event captured public interest and the story goes that it inspired New York shopkeeper Morris Michtom to display two toy bears made by his wife, Rose, in the window of his candy and stationery store, nicknaming them 'Teddy's Bears'. Taking inspiration from Berryman's innocent depiction of the bear, Rose Michtom's bears were sweet and cuddly, and had instant public appeal. Realising he was on to something, Morris Michtom went into production and founded the first Teddy Bear Manufacturing company in the US – 'The Ideal Novelty and Toy Company'.

President Theodore 'Teddy' Roosevelt, whose refusal to shoot a bear was satirized in a cartoon in the *Washington Post*. This in turn inspired the creation of the teddy bear.

At the same time, over in Germany, Richard Steiff (1877-1939), an artist and the nephew of Margarete Steiff, who ran a stuffed toy business, inspired his aunt's company to make its first prototype toy bear, based on

ORIGINAL FACTS

Did you know the technical term for someone who collects teddy bears is an 'Arctophile'?

his sketches. Two quite different types of bear therefore evolved. Steiff's with its long snout and humped back looked more realistic than Michtom's more wide-eyed cartoon-like bear. Neither the Steiffs nor the Michtoms knew about the other's bears until the Leipzig Toy Fair in 1903, when an American buyer, who was aware of the teddy bear fever in the States, bought 3,000. It was a case of Steiff being in the right place at the right time.

Today, teddy bears still enjoy immense popularity and have become collectors' items. In 1982, the first ever auction of teddy bears was held at Sotheby's in London. The press coverage of this sale caused waves of collecting fever throughout the world, and each successive auction saw higher and higher prices, until the present world record of £110,000, which was paid for 'Teddy Girl' at a sale at Christie's in South Kensington. This bear was the lifelong companion of the world's most famous bear collector, the late Colonel 'Bob' Henderson. 'Teddy Girl' now resides in a Japanese museum.

TENNIS

Tennis is derived from a medieval French game called *Jeu de Paume*, which means 'game of the palm'. This is somewhat paradoxical given that tennis is the racquet sport par excellence. The game is believed to have been first played by monks in the cloisters of French and Italian monasteries. They adapted the rules to suit the particular layout of their buildings in the much the same way as schoolchildren do. The laws of the game gradually became unified as monks travelled between monasteries with a desire to take on their rival brothers. They began to hit the balls with padded gloves instead of bare hands and later took to using wooden paddles. When a

A game of Real (or 'Royal') Tennis played at Windsor Castle circa 1500, in the reign of Henry VII. Note the slopping roof of the gallery on the left, onto which the balls were served.

monk was about to serve, he would cry out 'Tenez!' ('get ready!'), which is thought to be the origin of the sport's English name.

By the thirteenth century there were as many as 1,800 *Jeu de Paume* courts in France. The Pope and the Holy Roman Emperor Louis IV (1281-1347) became so alarmed by the popularity of the game that they tried, unsuccessfully, to ban it. By now *Jeu de Paume* had spread from the monasteries to the royal courts of Europe. According to Shakespeare, the French Dauphin riled Henry V ahead of the Battle of Agincourt in 1415 by sending him a mocking gift of tennis balls. The English king replied that he would turn them into cannon balls.

The stringed wooden racquet had been developed by 1500 but the sport was still a long way from its modern form. At this stage, the only version of the game was what is now known as Real Tennis ('Real' is the Spanish word for 'royal'). This featured a droopy net, an enclosed court equipped with buttresses and a gallery with a slanting roof onto which the balls were served. Lawn Tennis came into existence only in the second half of the nineteenth century. The catalyst was Charles Goodyear's (1800-1860) invention of vulcanised rubber, which allowed the manufacture of balls sufficiently bouncy to work on grass.

THERMOMETER

The earliest known thermometer was invented by Galileo Galilei (1564-1642) in about 1593. It consisted of a glass bulb that opened out into a long thin tube. When the tube was dipped in cooled water and the bulb was heated, the air inside the device expanded and some of it was expelled. When the source of heat was removed, the remaining air in the thermometer contracted and water was sucked up the tube. The quantity of water reabsorbed was an indication of the temperature of the original heat source.

Galileo's device was useful for comparative purposes but gave no 'absolute' readings. For that reason it was strictly speaking a thermoscope. It only became a thermometer when a numerical scale was added by Santorio Santorio (1561-1636) around 1612. The problem was that readings were still liable to be distorted by variations in atmospheric pressure. In 1714, the German physicist Daniel Fahrenheit (1686-1736) came up with a solution in the form of the sealed mercury thermometer. By heating the instrument ahead of sealing to expel all the air, the liquid metal was left free to move in a vacuum. This made external air pressure irrelevant.

Fahrenheit is generally thought to have defined 100°F by his body temperature on the day he took the decisive reading. Presumably he either had a slight fever or his thermometer was slightly inaccurate. The manner in which he arrived at 0° on his scale Is less certain. One theory holds that he used the lowest temperature he recorded during the winter of 1708-9 in his home town of Danzig (now Gdansk in Poland). Another explanation is that he derived his zero from the melting point of an equal mixture of salt and ice. In 1742, the Swedish astronomer Anders Celsius (1701-1744) devised a simpler scale based on the freezing and boiling points of water. Initially he defined the former as 100°C and the latter as 0°C, but this 'upside down' system was later reversed.

THERMOS®

The insulating properties of spaces devoid of air were understood by the mid-seventeenth century, but it proved very difficult to build equipment that maintained a vacuum. The first reliable Thermos flask – thermos means 'heat' in Greek – was made by the British physicist Sir James Dewar (1842-1923) in 1892. It consisted of two glass cylinders, one inside the other, hermetically sealed with an evacuated space between them. The inner cylinder was coated with silver to further reduce the transfer of heat.

In the early days, the use of Dewar flasks, as they were initially known, was restricted to the scientific community. They were chiefly employed

US Patent 872,795 3 December 1907

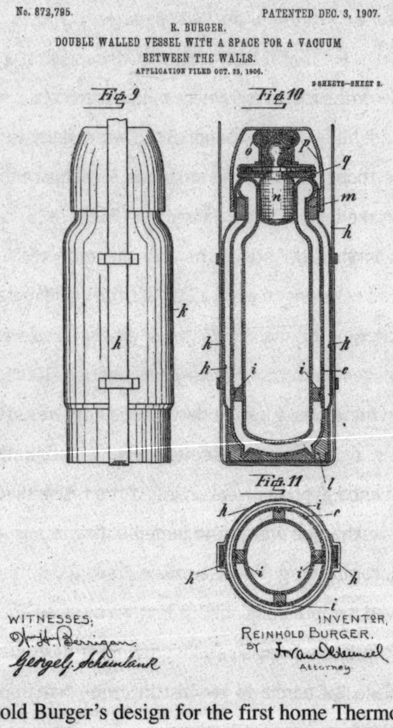

Reinhold Burger's design for the first home Thermos flask, or vacuum flask to give it its technical name.

to keep serums and vaccines at stable temperatures. But Reinhold Burger (1866-1954), the German glassblower who built the original flasks for Dewar, saw the potential of a home version. He modified the design to incorporate a protective metal exterior and was awarded a German patent in 1903. (Dewar had declined to patent his invention as he believed it should be freely available to scientists.) In 1906, an American businessman named William B Walker came across Burger's Thermoses during a trip to Berlin, bought up the patent rights and began to import them into the US. They were quickly snapped up by housewives, huntsmen and anglers. Thermoses were also popular with explorers, and received invaluable publicity when Shackleton and Peary used them on their respective expeditions to the North and South Poles.

THONG

The thong has become such a trendy piece of clothing that it even has a song named after it. The hit by R&B sensation Sisqo was a great success in 2000 and the words 'Thong, th-ong, thong, thong...' catapulted the Dru Hill frontman up into the top 10 of the US and the UK charts.

There are two types of thong; one is a piece of underwear, which provides back coverage, with a strip of fabric that is usually one-half inch to one-inch thick. The other is formerly known as the G-string and is in fact, very stringlike, offering only the bare minimum. It would seem that the idea of the thong was taken from the loincloth, which was worn as an undergarment thousands of years ago. The earliest loincloths were made of leather and Egyptians were amongst the first to wear them. The famous king Tutankhamun was found buried with more than a hundred loincloths in his tomb.

The thong was, not surprisingly, adopted by exotic dancers and performers in strip clubs. Such a reckless baring of flesh prompted New York Mayor Fiorello LaGuardia to request in 1939 that the city's exotic dancers dress in a more appropriate manner. The designer who can be credited for the modern thong is Rudi Gernreich (1922-1985), a fashion designer who was the first to introduce in 1974 the thong as we know it today. Gernreich was born in Austria in 1922 and worked in the fashion industry in the 1950s in America. He also invented the topless swimsuit or monokini, which became a couture emblem of the 1960s sexual revolution. Gernreich has therefore been described as much a futurist as a designer. The thong was enthusiastically adopted in Brazil, which probably has a lot to do with the country's 4,600 miles of coastline. The thong is known as the 'tanga' in Brazil.

The loincloth was the original inspiration for the thong.

The thong is now one of the fastest selling underwear styles available amongst women and is also becoming of interest to men. However, in Melbourne, Florida, it is very unlikely that you will spot any thongs; since January 2005, it has been illegal for anyone over the age of 10 to bare their buttocks in public; the punishment is a $500 dollar fine or 60 days in prison.

TIE

While scarves, shawls and neck-cloths can be traced back to ancient times, the modern necktie is a relatively recent invention. Its first appearance was in the mid 1600s, when a band of Croatian mercenaries visited Paris to pledge their support to King Louis XIV. Ever keen to start new fashion trends, the French aristocracy were quick to notice the distinctive neckwear sported by the Croatians, and it soon caught on; the expression 'à la Croat' corrupted to the familiar 'cravat'. Some sources dispute this, saying the word is more likely rooted from *rabat*, meaning a hanging collar.

Charles II brought the necktie to England soon afterwards, when he returned from exile to reclaim his throne after the Puritan revolution. Neckties soon became an indispensable part of a gentleman's wardrobe; the connoisseur of cravats, M Le Blanc, went so far as to say: 'The grossest insult that can be offered to a man... is to seize him by the cravat; in this place blood only can wash out the stain upon the honor of either party.'

The rise in popularity of the long tie that tapers towards the throat occurred largely in the nineteenth century, when the dandyish look began to give way to a more austere suit, closer to the modern business style and famously advocated by George Brummell. By the 1850s ties had begun to have another function, as signifiers of social groupings in the form of club or school ties. The four-in-hand also became the most widely favoured knot. In the 1920s, Joss Langsdorf invented the slip stitch, a loose black thread running through the ties which increases its resilience and helps the tie maintain its shape. In the 1960s and 1970s, ties became notably fatter, sometimes up to five inches wide, in proportion with larger collars and lapels.

Today, ties imply a certain formality and seriousness of intent, but are often used to inject a streak of colour and to suggest a spirit of individuality in the otherwise uniform world of big business. The enduring appeal of ties is perhaps best summed up by a remark from couture author and marketing expert, Michael Solomon: 'they are not even practical. Yet the tie remains an essential part of a man's wardrobe because it unites all the elements of a man's outfit, giving him instant respectability and, above all, It Is the ultimate symbol of individuality.'

Some knots and types of necktie:

- The Four-in-hand
- The Lord Byron
- The Windsor
- The Half-Windsor
- The Bow-tie
- The Cravat
- The Bolo tie
- The Clip-on tie

TOASTER

Bread has been a staple food for mankind for millennia. The habit of toasting it only appeared in Roman times, who no doubt discovered that, as well as giving it a crunchy flavour, toasting bread preserved it for longer.

The toasting process is pretty simple. When a bread slice is heated up to a temperature of around 310°F, a chemical change known as the Maillard reaction begins. Sugars and starches start to caramelise – turning brown – and the bread takes on intense, smoky, mouth-watering flavours.

For many centuries the only way of toasting bread was to rest a slice on a pan or a grill, or to prop it with all sorts of oddly-shaped metal pincers to hold it near the flame. With the introduction of electricity in the late 1800s, and the discovery in 1905 of Nichrome – an alloy of nickel and chromium – the first electric toaster was just a step away. Thomas Edison's (1847-1931) power could create heat without burning the place down, and the new metal helped Albert Marsh (1877-1944) to develop a temperature-resistant wire.

Over the following decade, several models were produced and the new invention became a common sight in many kitchens. In 1919, Charles Strite designed the pop-up toaster, and, in 1926, introduced the Toastmaster: an automatic toaster that could brown bread on both sides simultaneously; set the heating element on a timer; and eject the toast when finished. After that, its popularity surged. By the 1960s, it was widely available to most middle-class American families. Since then, the design has remained largely the same, but improved by a few innovations, such as the use of cheaper materials, heat-resistant plastic, variations in the size of the slots, allowing bagels and doorstep slices to be toasted too and the addition of defrosting and reheating functions. And in 2001, British design student, Robin Southgate, developed a toaster that took meteorological information from the internet and then browned the bread with an image of what weather to expect on the way to work: sunny, cloudy or rainy.

US Patent 1,394,450 18 October 1921

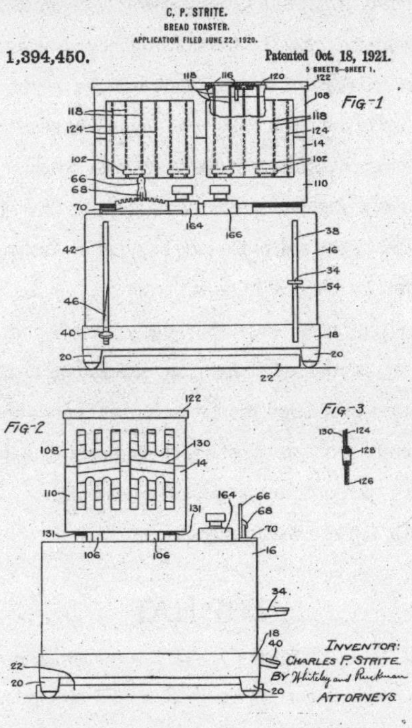

The first pop-up toaster, designed by Charles Strite.

TOOTHPASTE

Imagine brushing your pearly whites with ground eggshells, lizard livers or even urine. These questionable ingredients were integral to some of the earliest recipes for tooth cleaners, dating back to ancient Chinese and Greek civilisations. A popular Egyptian concoction of 5,000BC was made up of ox hooves, myrrh and pumice. And a powdery substance used to remove debris from Persian teeth in 1,000BC was a deeply unpalatable blend of burnt snail shells and gypsum.

Improvements were a long time coming; but in the eighteenth century the most widely accepted basis for toothpowder – or 'dentifrice' as it was widely known – was bicarbonate of soda, with Borax added to provide the foaming effect. In the early nineteenth century, chemists began to add the newly available binding agent, glycerine, transforming the powder into a thick, sticky paste. Further modifications came in 1824 from the dentists Peabody, who contributed soap to the mixture, and John Harris, who added chalk.

In 1873, Colgate was the first brand to mass-produce fresh smelling toothpaste in a jar. However, this packaging was short-lived and was replaced in 1892 by the collapsible tube brand of Dr Washington Sheffield, which he named Dr Sheffield's Creme Dentifrice.

The use of toothpaste increased hugely after World War II, when emulsifying agents such as Sodium Lauryl Sulphate displaced the less flavoursome soap. In addition, the discovery that fluoride helped to reduce tooth decay enshrined it as the star ingredient, ensuring that toothpaste at last had some proper scientific clout. The witch-doctor combinations of the past were a dim and distant memory.

TOP HAT

On the evening of 15 January 1797, a haberdasher named John Hetherington strolled out of his shop in the Strand in London and precipitated a riot. The cause was his headgear. Hetherington had modified a contemporary riding hat in his workshop, giving it a taller crown, narrower brim and a covering of black silk instead of beaver skin. The sight, according to the *London Times*, was enough to induce panic. Several women fainted and an errand boy's arm was broken in a melée of curious passers-by. Hetherington was arrested for disturbing the peace and fined £50 for wearing 'a tall structure having a shining lustre calculated to frighten timid people'. But he soon recouped his losses because the publicity surrounding the case led to a deluge of orders.

TRAFFIC LIGHT

Before the traffic light there was the traffic signal, invented by JP Knight in 1868. His lamppost-like device, raised in London at the junction of George and Bridge Streets near the Houses of Parliament, had two semaphore arms that would pop out to a horizontal position in order to halt a stream of oncoming traffic. A gas lantern was added to the top of the main pole, which glowed either red or green to enable the nighttime traffic to see it.

Unfortunately, the signal had certain limitations, in that it required a bystanding operator in order to work. One such operator – a member of the London constabulary – came a cropper on 2 January 1869, when the gas lantern on the prototype signal blew up and injured him.

US Patent 976,939 29 November 1910

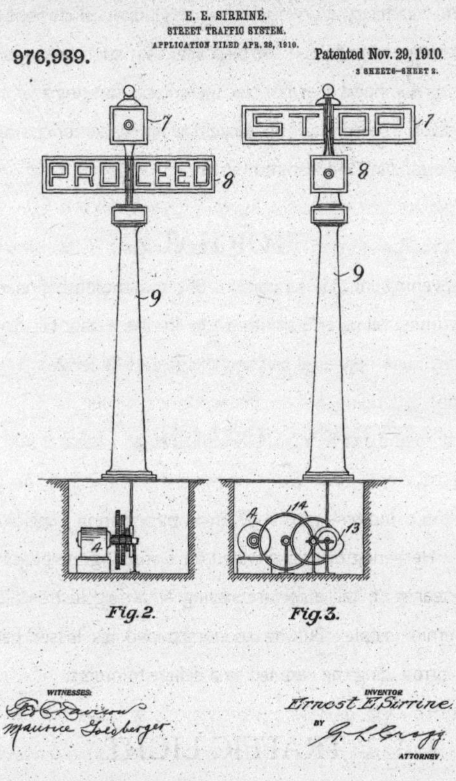

Earnest Sirrine's 1910 traffic light system.

As traffic became more automated, so did the signs, and American innovators took the lead. Chicago's Earnest Sirrine came up with a non-illuminated system using the words 'stop' and 'proceed' in 1910, then in 1912, Lester Wire (1887-1958) of Utah hit upon the first electric version with red and green lights. The following year, Cleveland's James Hoge patented a new set of manually controlled lights, which were later installed in Cleveland, Ohio. His lights used the words 'stop' and 'move'. Also of note is William Ghiglieri's 1917 signal with red and green lights that could be switched between manual and automatic. Then, sometime around 1920, Detroit policeman William Potts gave the world its first automatic, overhanging system with an additional yellow light to mark the transition from green to red and back again. Whether Potts was either too generous or too naïve to claim credit is unclear, but he never patented his innovation. It was left to the tenacious African-American businessman Garret Morgan (1877-1963) to officially document his creation of a cheap manual system in 1923, patenting it in the US, Great Britain and Canada. This became the standard method of traffic control for many years, until it was replaced by improved versions of the automatic, three-light set-up we know today. Of particular interest to New York fans, the first 'walk' and 'don't walk' signs appeared on the city's streets in 1952.

TRANSFORMERS TOY©

The Transformers franchise originated when the American toy manufacturer Hasbro – guided by the entrepreneurial Holocaust survivor, poker champ and multiple patent-holder Henry Orenstein (b1925) – met representatives of the Japanese company Takara at the Tokyo Toy Show in 1983, and proposed marketing the successful Diaclone and Micro Change toylines in the US under the new name Transformers.

The primary goal of Hasbro was always to sell toys, but they upped the ante by creating not only a cartoon and Marvel comic series (as well as a full-length film in 1986), but also an entire world and complex sci-fi back-story of their sentient robot heroes.

The premise – devised for Hasbro by Marvel writers, Jim Shooter (b1951) and Bob Budiansky – is basically a good old-fashioned tale of good versus evil: the honorable and virtuous Autobots are engaged in a seemingly endless battle with the cunning and downright evil Decepticons. Throughout the various overhauls and re-imaginings of the Transformer universe, this intergalactic war has spanned hundreds of thousands of years, taking place in many disparate locations, including pre-historic and contemporary Earth.

The initial run of products, referred to as Generation 1, lasted from 1984 to 1990 in the US; they fared better in Japan and the UK, where production continued until 1992. That said, Transformers toys have been made and sold every year since their release.

The basic Transformers mythology holds that the two factions, having virtually destroyed their home planet with their incessant warfare, fled into outer space, some of them crash-landing on Earth, where they remained dormant until the year 1984 when they were awakened to continue their conflict with earth as their battleground.

The success of Transformers in capturing the imagination of the kids of the 1980s lies chiefly in the gimmick their name suggests: they transform. The humanoid contraption can, with a certain amount of wrestling and consulting of the accompanying diagrams, be re-configured into at least one other shape – in the case of Generation 1 usually a car, truck or plane. They also came branded with a badge on the chest, which when vigorously rubbed would reveal whether your toy was an Autobot or the infinitely cooler Decepticon. A common complaint was that the toy never quite transformed in the well-oiled way it did on the TV series, and the manufacturers went some way to correcting this in the subsequent Beast Wars spin-off, in which a single part of the toy could trigger the entire transformation.

There is currently a Transformers revival gaining momentum, with news of a feature length film produced by Steven Spielberg to be released in 2007. And, like many toys of yesteryear, there is also a fan club, or in this case a profusion of Transformers fan sites, in which connoisseurs can compare notes on weapons capabilities and create their own screenplays and plots, or even, hilariously, character profiles. For example, this choice description of the evil Bonecrusher: 'Rare is the edifice that he feels is better off left standing. For him, demolition is not merely a job – it is a performance.'

TROUSERS

Trousers first popped up way back in the fourth century BC when both the Scythians and the Persians (men and women alike) adopted trousers for warmth and comfort, which suited their nomadic lifestyles. A little further away, King Wu of Zhao in 375BC, noting the style on foreign horsemen on his northern border, also decided to adopt the look for his own cavalry.

Trousers as we now know them evolved from the highly unattractive 'hose' that was worn in the fifteenth century. One hose for each leg, men would secure these to a buttoned jacket used as an undergarment, otherwise known as a 'doublet'. Gradually over time, both hose were joined at the back, then front, leaving a space for bodily functions. Doublets covered the knees, covering genitalia, but as fashions changed and doublets became shorter, it became necessary and enforceable by the church for men to cover up with a codpiece. Luckily for us all, by the end of the sixteenth century, breeches were Invented, incorporating the

A seventeenth century Englishman, wearing trousers far more like our own modern garments.

DID YOU KNOW?

- The word 'pants' is a derivative of 'pantaloons' which are ankle length trousers.
- As late as 1970, in some parts of the US it was illegal for women to wear trousers in classrooms, offices and restaurants.
- The word 'trousers' is ironically of Scottish origin, who are better known for their love of kilts than their trousers.

two hose and codpiece in one. Breeches were knee-length and came with their very own zip/button opening. Interestingly, this is why trousers are always plural, never singular. Then, during the French Revolution, men replaced knee-length breeches with a longer ankle-length style.

Trousers did not become commonplace attire for women until the late twentieth century. In the nineteenth century women were socially permitted to wear trousers for horseriding, but only hidden under heavy skirts. Screen icons such as Marlene Dietrich (1901-1992) and Katharine Hepburn (1907-2003) were photographed in trousers during the 1930s in a bid to make them more acceptable for women, but it was quite shocking even then. During World War II women could wear fitted men's trousers to work in factories, and after the war it became acceptable for women to wear them for leisurely pursuits such as gardening and beach wear. However, the enormous popularity of jeans changed all this for ever. Jeans united the masses in a sea of blue denim regardless of gender and class. Jeans were also made popular by sailors, who were one of the first to sport the look, due to their durability.

TUPPERWARE®

Tupperware is now a worldwide phenomenon, with net annual sales exceeding US$1.2 billion, and is famously used even by the queen of England. It was invented by Earl Silas Tupper (1908-1983) from New

Hampshire, USA. Tupper started out as a tree surgeon, but later took a job at the Du-Pont Chemical factory in Massachusetts, which had been developing plastic before World War II. He only worked there a year, but he claimed this was formative in developing his understanding of the design, research and development of plastic. An inventive man by nature, he used this knowledge to start up his own factory in 1938. The Earl S Tupper company initially sub-contracted work from the Du-Pont factory, but during the war it began to develop gas masks and other items to help the war effort.

It was after the war that Tupper began to focus his attention on plastic consumer products. This was a completely different market to the war effort. The plastics industry was still relatively primitive and the products and materials available were not up to standard for the consumer market. He began by inventing a method to transform polyethylene slag, an unattractive by-product of the crude oil refinement process. His experiments were a significant step forward in the plastics industry and allowed him to make lightweight, non-breakable containers, cups, bowls and plates. He later developed leak-proof air-tight lids, by inverting the design of the paint tin lid.

However, Tupperware was not an overnight success. Consumers were initially sceptical about this new product. It was not until the late 1940s that sales began to take off, when Tupper began to realise the potential of door-to-door sales. By 1951 he had removed Tupperware from store shelves and was purely selling through direct home sales. Salesmen introduced their products to householders assembled at a hostess's home. The result was 'Tupperware parties' which are still going strong today. Today a Tupperware party begins somewhere in the world approximately every two seconds. It seems some people can never have enough storage.

TV COMMERCIAL

The world's first television commercial was broadcast by WNBT-TV in New York City at 2.29pm on 1 July 1941. It was an advertisement for

Bulova watches consisting of one of the timepieces displayed over a map of the USA with the voiceover 'America runs on Bulova time!' The manufacturers paid US$9 for the 10-second slot, which was screened before a baseball match between the Brooklyn Dodgers and the Philadelphia Phillies. The British equivalent, a commercial for 'tingling' Gibbs SR toothpaste, was broadcast on ITV at 8.12pm on 22 September 1955.

U

ULTRASUEDE®

Ultrasuede looks and feels like soft leather but is man-made, washable, durable, never has to be ironed, and is resistant to water. It is relatively expensive – a metre of fabric costs upwards of $100 (around £60), but is so hardwearing that an item of clothing should last a lifetime.

Ultrasuede is a trademark owned by the Toray company that invented it in 1970. After years of trying, Dr Miyoshi Okamoto of Japan invented the first ultra-microfibre, but it was his colleague, Toyohiko Hikota, who transformed the ultra-microfibre into Ultrasuede a few months later, using a 'bicomponent spinning technique'.

By combining two different polymers – micro-polyester with polyurethane – he created a new material that had all the qualities of suede but also the creaseproof nature of a polyester shirt. While rival American companies were trying to imitate natural products, the Japanese companies were looking to invent fabrics that surpassed natural ones in quality, longevity and, of course, price. Toray was originally looking for a substitute for silk, but stumbled across this pseudo-suede instead.

The background of microfibre research shows that DuPont – the company that invented nylon – had the technological know-how to develop this method of spinning in 1964, six years earlier, but did not have the foresight to see its developmental potential. This meant that the Japanese were identified as the market leader.

Okamoto was able to develop his ultra-microfibre due to the 'angura' policy at Toray, which stated that workers were allowed to spend 20% of their time working on their own entrepreneurial ideas that only they and their supervisors knew about. This policy worked well for Toray whose Ultrasuede products have been flying off the racks ever since. Microfibres are so-called because they are so thin, less than one denier, (nylon stockings are usually 15-20 denier), and 100 times thinner than a hair from your head. It is claimed that one pound of the fabric could stretch to the moon and back.

UMBRELLA

The word 'umbrella' derives from the Latin *umbra*, meaning 'shade'. As this suggests, umbrellas were originally employed not to keep their owners dry but to protect them from the sun. Such devices are also known as parasols. They were depicted in bas-reliefs (sculpted wall surfaces) in Nineveh in Assyria around 1,400BC and began to appear in Egyptian wall carvings shortly afterwards. The Assyrian umbrella appears to have been reserved for royalty, but the Egyptian equivalent was used by ordinary citizens as well as some of the gods. It was associated with the sky goddess Nut, who was herself understood as a kind of protective cosmic umbrella. The folding variety seems to have been invented in about 300AD in China. The Chinese were also the first to make their umbrellas waterproof, which they did using lacquer.

The Ancient Greeks and Romans also possessed umbrellas, but their use was largely restricted to women. With the notable exception of the Pope, this remained the case in Europe until the mid-eighteenth century. The first Englishman with the temerity to carry an umbrella in public was an

Umbrellas have been around since at least 1,400BC, though they were initially used to protect people from the sun, not the rain, much as today's parasols do.

eccentric traveller named Jonas Hanway (1712-86), who had picked up the habit in Persia. He was constantly ridiculed, particularly by coachmen, who saw the umbrella as a potential threat to their rain-influenced trade. According to the contemporary author John MacDonald, whenever they saw Hanway brandishing a brolly, they would shout, 'Frenchman, Frenchman, why don't you call a coach?' Eventually, however, his persistence won the day. By the time Hanway died, the umbrella was well on its way to becoming a standard accessory for the gentlemen of London.

UNDERTAKERS

The first recorded undertakers in history were Egyptian priests of the god Osiris. They were charged with the solemn task of ensuring the safe passage of the souls of the deceased to the afterlife. The spirit or ka was believed to be an exact double of the dead body which would only be accepted by the gods if the latter was kept intact. This was the rationale behind the seventy day mummification process. As only the very wealthy could afford the procedure the poor were effectively denied access to the hereafter.

The first stage of embalming involved the removal of the internal organs. The priests removed the brain first, via the nostrils, and then the liver, lungs, stomach and intestines were dried, wrapped and placed in special containers called canopic jars. Only the heart was left inside the body as it was believed that it would be weighed against the feather of Ma'at, the goddess of justice, to decide whether the ka qualified for admission to the domain of Osiris.

The first recorded undertakers are Egyptian priests, who performed the seventy day mummification process believed necessary to ensure the deceased's entry into the afterlife.

The corpse was then filled and covered with a form of salt called natron and left for 40 days to dry out. It was then washed, packed with linen and wrapped in hundreds of yards of bandages. If all the correct spells and incantations had been recited, the deceased stood a good chance of immortality.

V

VACUUM CLEANER

If you think vacuum cleaners are noisy today, you might change your mind upon hearing Hubert Cecil Booth's (1871-1955) invention of 1901, the Puffing Billy. A horse-drawn contraption the size of a car (also fuel-driven and parked outside the building), this dust-busting monstrosity was to blame for frequent lawsuits – the volume was just too much for the frightened horses passing along the road. The machine's long hoses had to be fed through windows and moved from house to house by a team of cleaners. Not the most convenient device, but truly revolutionary compared to the method that it replaced, of beating rugs with a stick.

In 1902, King Edward VII (1841-1910) decided that the state of the carpet in Westminster Abbey was unacceptable for his coronation, and that Booth's machine was his only hope. After ordering a demonstration of its capabilities, the king bought two cleaners, which, needless to say, secured Booth's success. The machine also made an appearance at the Royal Mint, though trust was not easily granted this time around. Upon his departure Booth was halted by the police – apparently the dust bag contained a large amount of gold dust, property of the Mint! The Puffing Billy was also spotted at London's Crystal Palace, whose girders required a good cleaning – the machine removed 26 tons of dust.

Before Billy, there had been other dust-busting pioneers. The 'Whirlwind' was a wooden machine powered by a suction fan, which had to be cranked by hand. The first portable electric device was invented in 1905 by the American firm Chapman and Skinner, although it weighed 92 pounds and had a fan that was 18 inches in diameter.

In 1907 James Murray Spangler, a janitor who believed the carpet sweeper he was using was to blame for his asthmatic cough, set out to re-design it and put an end to his allergies. Using an old fan motor, a soapbox, a broom handle and a pillowcase to collect dust, Spangler created a makeshift prototype of an electric powered vacuum cleaner. He improved the model using ingredients other than household items, and received a patent for it in 1908. One of his first customers was a cousin who was married to William H

Puffing Billy – somewhat clunkier than a Dyson, but a revelation nonetheless when it appeared in 1901.

Hoover (1849-1932), the owner of a leather goods company. Having confidence in the product as well as the financial resources, Hoover purchased the patent from Spangler and created the Hoover Company. A few modifications were made and the first 'Hoover' model resembled a bagpipe attached to a large box. Sales were disappointing at first, so Hoover offered a 10-day free trial to prospective buyers. It wasn't long before it became a household name, and we were all doing the 'hoovering'.

Vacuum Folklore:

● In 1993, one Australian door-to-door salesman piled a mountain of dirt onto a couple's beautiful new carpet. When his machine failed to remove it properly, the man was subsequently attacked by the couple and ended up in hospital.

● A common sales pitch for travelling salesmen involved asking the 'lady of the house' to remove the linen and covers from her bed. Next, he would open his cleaner's dirt bag and pour some of the contents onto it. After announcing to her that 75-80% of household dirt is comprised of human skin, and that she had practically been living in it, a sale was nearly always guaranteed!

VCR (VIDEO CASSETTE RECORDER)

The first video tape recorder (VTR as it was then known) was invented in the early 1950s and cost an astonishing $50,000. Consequently, it was only television companies who could afford it, but it earned the Ampex Corporation an Emmy in 1957 for inventing it. Ampex had brought out the audiocassette in 1946 and wanted to progress to video recording. They hired a man named Charles Ginsburg (1920-1992) who achieved this breakthrough by speeding up the magnetic tape used for audio recording to enable the recording of higher frequency video. The VTR meant that television companies could pre-record and edit their programs in a cheap and speedy way; CBS were the first to do this in 1956, when they broadcast the news two hours after recording it. Ginsburg's machine was hailed as 'one of the most significant technological advances to affect broadcasting and program production since the beginning of television itself'.

The race to invent the domestic video recorder started in the 1960s, and in 1963 the Nottingham Electric Valve Company released Telcan, a machine that recorded reel to reel, although it was complicated to operate. A far more promising machine was Cartrivision, brought out in 1972. An all-in-one TV and video recorder, it was released along with a whole host of pre-recorded films. However, when the tapes it used were found to have disintegrated while stored in a warehouse, investors became nervous and Cartrivision was canned.

Philips brought out their video recorder in 1972 and coined the generic term VCR. The catchily-named N1500 model was similar to Sony's U-matic, brought out the previous year, but was less expensive and aimed at the domestic market. Yet at £600 a pop (equivalent to £4,500 in today's money), it was an awkward sell, and problems with pricey tapes that tore easily meant that there was still a long way to go.

Sony decided to try to enter the home recording market with the ill-fated Betamax. Released in Japan in 1975, Betamax was a bestseller by 1983 due to its superior quality. Sony then tried to standardise the videocassette so that no one was excluded from the race, but their rivals

US Patent 2,956,114 11 Ocotober 1960

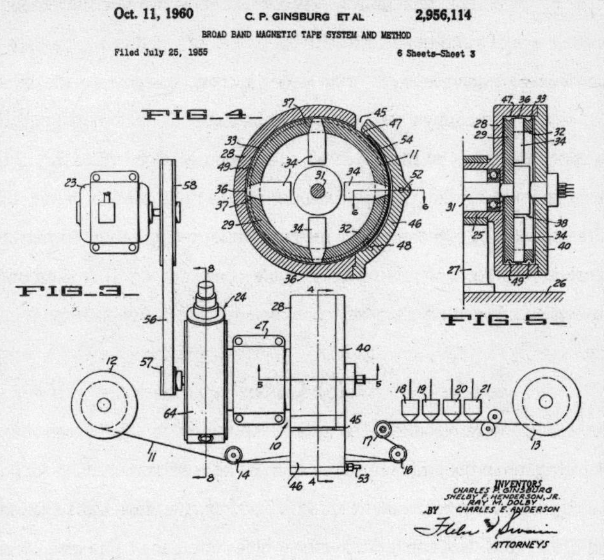

Charles Ginsburg's patent for the first VCR, hailed as 'one of the most signficant technological advances to affect broadcasting and programme production since the beginning of television itself'.

were not interested. Video rental shops had to have separate shelves for the Beta videos and the VHS videos. Slowly VHS overtook the shop while the racks for Beta became smaller, until eventually they disappeared altogether. Sony had to admit defeat by 1988 and started producing VHS tapes instead.

The reasons for the failure of Betamax are numerous: the machines and tapes were more expensive, and yet the difference in quality was almost imperceptible to the layman. The Beta tapes would not record for as long as the VHS tapes, and a one-hour tape was useless if you wanted to record a feature film. Beta did bring out a longer tape, which traded quality for length, but by then it was too late. Also, while VHS machines could be rented at all major high street stores, Beta had to be bought; it was thought of as the 'connoisseur's choice', so Sony thought that enough people would make the investment. They were wrong. It is also thought – somewhat mischievously – that VHS's success was due to the sheer volume of pornography available on the format.

Since the debut of DVD in 1996, the prognosis of the VCR has looked very poor indeed. DVD machine sales outpaced those of VCRs in the early 2000s, and in 2006 the Video Software Dealers Association announced that no more major film releases would come out on VHS. Ginsburg's spectacular invention is unlikely to witness its own 100th birthday.

VELCRO®

In 1948, Swiss engineer George de Mestral (1907-1990) was on a hunting trip in the Jura Mountains when he noticed that his dog's fur and his trousers were becoming covered in cockleburs – loose seeds enclosed in furry husks that catch onto animal hides in order to find new places to colonise. Fascinated by their clinging capabilities, de Mestral used his home microscope to examine the interplay between the seeds and his clothing. He found that the seed husks were covered in tiny hooks that snagged on the loops standing out from the fabric of his clothes. It was a 'eureka moment'.

Over the next few years, de Mestral, using synthetics, experimented with a huge range of different versions of what he had envisaged as a new kind of fastener. After much hard work, trial and error, he reached his goal with a stroke of inspiration; he worked out that when nylon is sewn under infrared light, it develops into tough, hardwearing hooks that can link up with loops in a corresponding surface of fuzzy, man-made fibre.

Taking the 'vel' from velvet and the 'cro' from crochet – the French for hooks – de Mestral gave his invention a name and entered manufacturing history. The Velcro SA company was launched in 1952, and in 1955 de Mestral registered a Swiss patent for his creation, entitled 'Velvet Type Fabric and Method of Producing Same'. In 1957, American Velcro opened

US Patent 2,717,437 13 September 1955

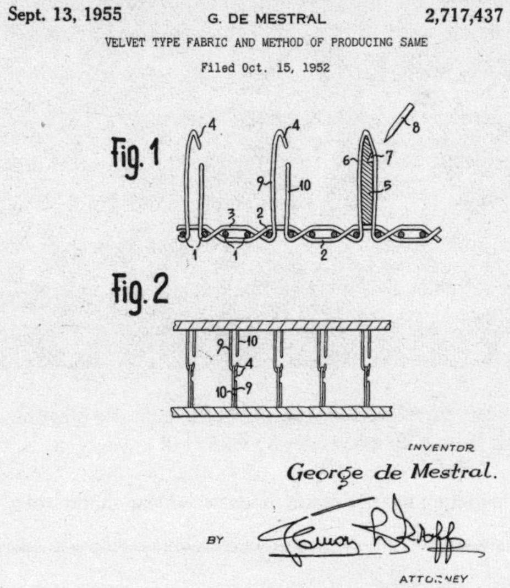

The hook and loop system used by de Mestral for Velcro is modelled on the way cockleburs snagged themselves onto the loops in the fabric of his clothes.

for business, and the following year saw the registration of the fastener's US patent. Soon after, Velcro was being produced at a rate of 60 million yards a year and was immortalised in 1985 by ZZ Top's song, 'Velcro Fly', which kindly reminded us that, 'You need just enough of that sticky stuff/ To hold the seams of your fine blue jeans'.

VENDING MACHINE

It was the Greek mathematician, Hero of Alexander, who in 215BC invented a machine for doling out measures of votive water in sacred Egyptian temples in exchange for a coin or two. He describes his creation in his collection of writings, *A Treatise on Pneumatics*, in which he elaborates upon his many other feats of engineering, including a steam turbine.

Japan has embraced the vending machine like no other country, filling them with a bewildering variety of unlikely items, from beer and rice all the way to – disturbingly – used underwear, although the police have clamped down on the latter.

However, the development of the modern vending machine had to wait until the 1880s, and began with a postcard vendor that was launched on the streets of London early in the decade: brainchild of renowned coin-op godfather, Percival Everitt. Around the same time, Richard Carlisle produced a machine for dispensing books; then, in 1888, the Thomas Adams Gum Company unleashed its army of tutti-frutti gum dispensers that marched across the US like brightly coloured robots and took up places in convenient alcoves from coast to coast. Gumball dispensers arrived in 1907, and William Rowe produced a cigarette machine in 1926.

Once the vending machine concept was established, subsequent innovators were pretty much free to fill them with whatever their hearts desired. The Germans invented the hot-dog vending machine, and the Japanese launched machines selling a cornucopia of unlikely items such as beer, rice, fresh vegetables and video games. The Japanese also invented an egg dispenser and a fast-food combo machine from which you could order French fries, burgers, breaded chicken, rice balls, sea bream and octopus.

On a less savoury note, Japan has also used vending machines to sell hardcore pornography and – as reported in *The Economist* in October 1993 – of used underwear, purportedly reclaimed from Japanese schoolgirls. In the latter case, worried police initiated a crackdown on the 90 or so machines dotted around the outskirts of Tokyo after finding that the entrepreneurs behind them had violated the national Antiques Dealing Law.

Japan has also sanctioned a Catch Your Own Live Lobster machine, placing the fee-payer in control of a large, three-pronged metal claw with which to terrorise a community of crustaceans cowering in a holding tank below. As if that were not enough, dispensers have proliferated in Buddhist and Shinto temples, allowing visitors to purchase stickers for decorating votive plaques, or soft drinks to calm them prior to worship, which takes the vending machine back to its devotional roots.

VIDEO GAMES

The video games industry owes its existence, and considerable wealth, to the efforts of some pioneering computer students at the Massachusetts Institute of Technology. Back then, these men were known as 'hackers': not programmers who hacked into other people's programs, but 'hacks' in an almost literary sense, cranking out ream after ream of primordial computer code. It was these hypergeeks who wrote the electric soup from which all subsequent computer programmes crawled.

But before they devised any serious uses for computers, such as banking systems or military defence, the hacks concentrated on having fun. In 1962, Steve Russell, Alan Kotok (b1941), Peter Samson (b1941) and Dan Edwards devised the first ever video game – an interstellar combat simulator called Spacewar.

'We had this brand new PDP-1,' Russell told *Rolling Stone* in 1972. 'It was the first minicomputer, ridiculously inexpensive for its time. And it was just sitting there. It had a console typewriter that worked right, which was rare, and a paper tape reader and a cathode ray tube display. Somebody had built some little pattern-generating programs, which made interesting patterns like a kaleidoscope. Not a very good demonstration. Here was this display that could do all sorts of good things! So we started talking about it, figuring what would be interesting displays. We decided that probably you could make a two-dimensional maneuvering sort of thing, and decided that naturally the obvious thing to do was spaceships.'

As video games are often blamed for ruining children's literacy, it is ironic that Russell drew inspiration from books in order to get Spacewar off the ground.

'I had just finished reading Doc Smith's Lensman series. He was some sort of scientist but he wrote this really dashing brand of science fiction. His heroes had a strong tendency to get pursued by the villain across the galaxy and have to invent their way out of their problem while they were being pursued. That sort of action was the thing that suggested Spacewar. He had some very glowing descriptions of spaceship encounters and space fleet maneuvers. By picking a world which people weren't familiar with, we could alter a number of parameters of the world in the interests of making a good game and of making it possible to get it onto a computer.'

As a pure concept, Spacewar drew huge acclaim from computer experts for what it enabled the early programmers to achieve. In a 1971 issue of Analog, Albert Kuhfeld wrote: 'Students were working their hearts out improving it, and the faculty was nodding benignly as they watched the students learning computer theory faster and more painlessly than they'd ever seen before.'

Spacewar was followed by Atari's Pong in 1972, Space Invaders in 1978, Pac-Man in 1980 and Tetris in 1985. Simple though they were, they paved the way for the explosion in computer games that was to follow.

W

WALKMAN®

Andreas Pavel (b1945) spent a quarter of a century battling with media company Sony for the legal right to call himself the inventor of the Walkman. Pavel – a well-travelled and artistic intellectual, born in Germany in 1945 and raised in Brazil – was a free-thinking philosopher and music fan who, inspired by long nights listening to records with friends in the late 1960s, dreamt up a device that would let him take his favourite tunes wherever he wanted.

With an industrialist father, an artist-musician mother and a home studio to play with, his family life provided the ideal conditions for Pavel's invention to take shape. Following periods of work in Brazil, where he worked in television and publishing, Pavel returned to Europe in protest at Brazil's military dictatorship. It was around that time, circa 1972, that he created his prototype 'Stereobelt'. Built in an L-shaped design, with a cassette-tape slot on one end and controls scattered over its matte-black surface, the Stereobelt was a clear precursor to what later emerged from Sony's factories. In an interview with the *International Herald Tribune*, Pavel recalled his first test of the device while in Switzerland, in the company of his girlfriend: 'I was in the woods in St Moritz, in the mountains, the snow was falling down, I pressed the button and suddenly we were floating. It was an incredible feeling, to realise that I now had the means to multiply the aesthetic potential of any situation.'

Delighted with his device, Pavel began to hawk it around to firms such as Philips, Grundig and Yamaha. But they failed to see its potential. 'They all said they didn't think people would be so crazy as to run around with headphones – that this is just a gadget, a useless gadget of a crazy nut,' said Pavel.

Undaunted, Pavel filed for Stereobelt patents in several different countries in 1977. This proved to be his masterstroke, as the range of filings would come to serve as a supporting network for future legal action. Then in 1979 came the bitter blow: Sony unleashed the Walkman. Within two years, Sony's trinket had shifted 200 million units and the Walkman's inventor was sold to the public as being Sony founder Akio Morita (1921-1999).

US Patent 4,412,106 25 October 1983

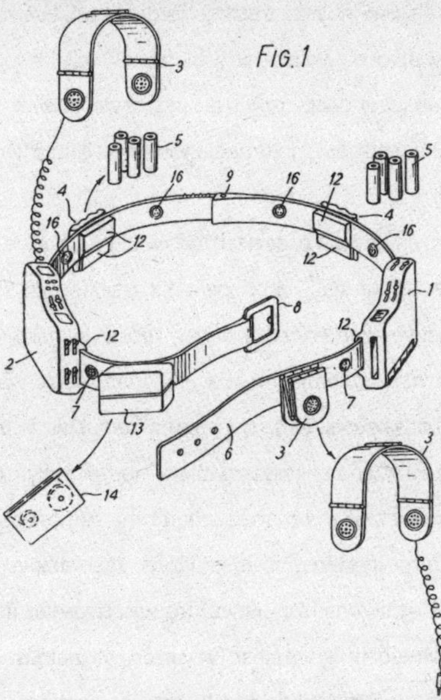

Andreas Pavel's US Stereobelt patent. He also had patents in numerous other countries, a fact which proved decisive when it came to winning royalties from Sony – not to mention acknowledgement as the Walkman's creator.

Pavel immediately initiated a gruelling series of lawsuits that would trail over the next two and a half decades, leaving him at one stage nearly bankrupt and facing court costs of more than $3 million. Just when it seemed that he would be crushed by Sony's legal machine, Pavel referred to his multi-national filings of the late 1970s and said that he would continue to sue Sony in every country that had granted Stereobelt a patent. Seeing a lengthy PR minefield ahead of them, Sony's legal team changed tack and sought to placate the determined Pavel. Fresh negotiations began in 2001 and, three years later, Pavel at last won the right to be recognised as the Walkman's inventor.

An out-of-court settlement rumoured to be in the region of $10 million may have gone some way to clearing his overdraft, but Pavel is reluctant to use his legal success as a platform for launching suits against makers of the Walkman's descendents, MP-3 players. Instead, he is focusing on his next creation – a personal multimedia tool he calls the 'Dreamkit'.

WASHING MACHINE

Catherine Beecher, an early proponent for transforming housework into an orderly and dignified enterprise, called laundry 'the American housekeeper's hardest problem'. For just one wash, boiling and rinse, roughly 50 gallons needed to be transferred from a pump or well to a stove and a tub. Then after the washing there was the drying, and then the ironing, a daunting task to say the least. It is no surprise that after the first washing machine appeared in the 1850s, the idea proved extremely popular. In 1851, the American James King was given the first of over 2,000 patents that would be granted to inventors of various clothes washing devices during the nineteenth and twentieth centuries. His was a hand powered, rotating cylinder device, the first ever to use a drum. One early model was driven by the 'unwinding' of twisted ropes, like a model airplane gains momentum from rubber bands. Another was called the 'Locomotive', named after its style – it moved back and forth along a track and washed clothes by slamming them against the walls of the tub. There were lots of creative ideas, but very few succeeded. Only two systems, the agitator and the tumbler, are still used today.

The first electric washing machine dates back to 1906. Often mistakenly credited to AJ Fisher, who patented a later drive, its true inventor is unknown. At the time it was quite a feat to create a water-based machine running on electricity, but even so this first effort was far from perfect. The motor was not properly protected, and water was able to drip into it and cause short-circuits. There were many challenges to creating a functional electrical washer: the motor needed to have sufficient starting torque, the power it produced needed to be transferred to the mechanism itself, and last, but

Hardly what we're used to nowadays, but this early washing machine was still a welcome innovation, and introduced the idea of a rotating drum still used today.

certainly not least, the danger of electric shock had to be prevented. To solve these problems, the machines were driven by chains, belts and gears, and a built-in fan prevented overheating and electric shocks. All this accomplished, and there was still one more consideration – how to prevent the clothes from shredding. In 1957, General Electric introduced a machine with a five-button control panel to modify agitation and spin speeds in addition to washing and rinsing temperatures. At long last, laundry had evolved from strenuous labour to the mere push of a button.

WATER WINGS

The inflatable plastic arm bands worn by children learning to swim were invented by Bernhard Markwitz of Hamburg. He was motivated to develop an alternative to the cork rings then in vogue by an incident in 1956 in which his three-year-old daughter nearly drowned. Fate lent Markwitz's researches a helping hand by supplying him with a DM 253,000 lottery win. He took the first two letters of his first and last names and formed a company called BEMA. After several years of development, BEMA water wings went on sale in 1964. The firm has since sold more than 150 million pairs.

WEATHER FORECASTS

People have been forecasting the weather since time immemorial. Jesus, for example, was well aware of the truth behind the proverb, 'Red sky at night, shepherd's delight, red sky in the morning, shepherd's warning'. In Matthew 16:2-3 he is recorded as berating the Pharisees and Sadducees with the following words: 'When evening comes, you say "it will be fair weather, for the sky is red," and in the morning, "Today it will be stormy, for the sky is red and lowering." You know how to interpret the face of the heaven, but you cannot interpret the signs of the times.' Other notable meteorologists of the ancient world include Aristotle, who wrote a treatise on weather called the Meteorologica in about 340BC, and the Babylonians, who were forecasting on the basis of cloud formations at least three centuries earlier.

The launch of a sounding balloon carrying a little box of weather instruments and radio equipment called a radiosonde.

The key to modern weather forecasting is the collection of readings taken simultaneously at locations spread over a wide geographical area. This so-called 'synoptic' system was made possible by the invention of the telegraph in the mid-nineteenth century. The first weather maps showing isobars (regions with equal atmospheric pressure) were made during the 1860s. The next great breakthrough was the development of radiosondes during the 1930s. These small boxes contain weather instruments together with radio equipment to transmit their readings to ground stations. They are born aloft by hydrogen or helium filled 'sounding balloons' and are able to accumulate atmospheric data to an altitude of 20 miles.

The world's first television weather forecast was transmitted by the BBC at 4.01pm on 20 November 1936. It lasted for six minutes. The American equivalent was broadcast on 14 October 1941 by the New York City station WNBT-TV (later WNBC). The bulletin was introduced by a cartoon character named Woolly Lamb with the immortal lines: 'It's hot. It's cold. It's rain. It's fair. It's all mixed up together. But I, as Botany's Woolly Lamb, predict tomorrow's weather.'

WEDDING RINGS

Couples are thought to have exchanged rings as part of Ancient Egyptian marriage ceremonies as early as the Third Dynasty (c2,800BC). The Egyptians invested the circle with great symbolism as it was the shape of both the moon and the sun. The latter in particular was identified with the supreme deity. Rings also signified eternity due to their endlessness. They were worn on the fourth digit of the left hand because this was believed to be the starting point of the so-called 'vein of love' thought to run directly to the heart.

WELLINGTON BOOTS

Although Wellington boots are now associated with farmers and wealthy urbanites seeking to prove that they own property in the country, they first came into existence in a military context. The boot was designed for Arthur Wellesley (1769-1852), the first Duke of Wellington, during the Napoleonic wars. Wellington found the boots then worn in the British Army cumbersome and uncomfortable, so he

The Battle of Waterloo won in a pair of wellies? It's highly likely, actually, as it was during the Napoleonic wars that the Duke of Wellington requested the first calfskin 'Wellington' boots, as an alternative to the boots worn by the British Army.

instructed his shoemaker, Hoby of St James Street in London, to come up with a better alternative. The resulting soft calf skin boot was closer fitting than its predecessor and lacked the latter's folded over top. As difficult as it is to picture the Iron Duke sporting a pair of 'Wellies' at the Battle of Waterloo in 1815, it is very likely that he did. He was certainly wearing the boots when he sat for a portrait by James Lonsdale during the same year.

Rubber Wellington boots were introduced by Hiram Hutchinson (1808-1869), who began selling them to French farmers in 1853.

THE WHEEL

Archaeological evidence suggests that the wheel was used to make pottery before it was adapted for transportation and milling purposes. Mesopotamian pots dating from the fifth millennium BC show signs of having been 'thrown' (ie shaped on revolving wheels) rather than coiled, but this is inference rather than proof. The earliest unambiguous image of a potter's wheel appears on a 5,500-year-old clay tablet excavated at the Mesopotamian city of Ur in Iraq. Ur was coincidentally the birthplace of the Old Testament patriarch Abraham.

The earliest depiction of a wheeled vehicle was incised around 4,000BC on the so-called Bronocice Pot in Southern Poland. Interestingly, the image is of a wagon with four wheels and two axles rather than a simple two wheeled cart. This suggests that wheel technology developed extremely rapidly once the penny had dropped.

WHEEL CLAMP/'DENVER BOOT'

As implied by the American name for the immobilisation device that is the bane of tardy motorists, the wheel clamp was invented by a resident of Denver, Colorado. It was patented in 1953 by Frank Marugg, a pattern maker and violinist for the Denver Symphony Orchestra. Marugg was an ingenious fellow who had built several interesting contraptions, including his own violin. He was also a friend of the city's sheriff, who asked him to design a device to immobilise the vehicles of residents who failed to pay their parking fines.

WINDSURFING

In 1968, a pair of southern Californian surfing enthusiasts named Hoyle Schweitzer (b1933) and Jim Drake (b1929) patented an articulated sailboard called the Windsurfer. During the following decade they were challenged by Newman Darby from Pennsylvania, who claimed to have built a sailboard as early as 1959. He had certainly written an article in 1965 describing a key component of the modern windsurfing board called the universal joint, a fact acknowledged by Schweitzer and Drake in their patent application. Ultimately, however, the courts rejected both claims in favour of an Englishman named Peter Chilvers (b1945). He was able to prove that he had constructed a board with a sail in 1958 and sailed it off Hayling Island in the south of England. At the time Chivers was 12 years old.

WONDERBRA™

In 1964, a lingerie designer called Louise Poirier developed a new bra for the Canadian lingerie company Canadelle. Although it had less padding than standard bras, the Wonderbra was capable of generating an impressive cleavage even when nature had failed to be generous in that department. The secret was its 54 design elements, which combined to lift the breasts and push them together without compressing them.

Soon after the invention, Candelle licensed the British manufacturer Courtland Textiles to sell the Wonderbra in Europe under its Gossard brand. It was launched in 1968 with the slogan 'Makes 34 like 36, makes 36 look pow!' Despite this promise, sales were unspectacular until the early 1990s, when the efforts of Madonna and a curvier new breed of supermodel brought the cleavage back into fashion. Things really took off in 1994 when Gossard launched an advertisement which featured the Czech model Eva Herzigova gazing down at her Wonderbra-supported chest in delight. The caption read 'Hello Boys'. The campaign was a sensation and was boosted considerably when billboard versions of the ad were blamed for several traffic accidents on the grounds of their hypnotic effects on male drivers.

Unfortunately for Gossard, its Wonderbra licence expired just at the point when sales went through the roof. The Sara Lee Corporation, which had purchased Canadelle in 1991, theoretically regained control of the brand on 1 January 1994. On 9 May, Sara Lee launched its American campaign by suspending a 2,800 square foot image of a Wonderbra-clad Herzigova over Times Square in New York City. The US captions were less suggestive than their British equivalents, but they had much the same effect. Soon Wonderbras were selling at the rate of one every 15 seconds.

WRIST WATCH

The world's first commercial wrist watches were manufactured by the Swiss firm Patek Phillippe & Co in 1868. There was, however, a non-commercial precedent. In 1810, Napoleon Bonaparte's sister Queen Caroline of Naples commissioned Abraham Louis Breguet to make her a 'bracelet watch'. Two years later he presented her with an exquisite wrist watch with a built-in thermometer. Wrist watches were considered exclusively feminine prior to the early twentieth century, as men preferred to carry pocket watches. But in 1904, Brazilian inventor Alberto Santos-Dumont (1873-1932) changed all that. He was building an aeroplane and wanted an easily visible timepiece, so he asked his friend Louis Cartier (1875-1942) to make him a 'manly' wrist watch, which he did – and a lucrative industry was born.

X

X-RAYS

The father of the X-ray was a German physicist named Wilhelm Conrad Roentgen (1845-1923). He made his fateful discovery on 8 November 1895 in a darkened laboratory in Wurzburg. At the time, Roentgen was investigating the effects of discharging an electrical current in a near vacuum inside a glass device called a 'Crookes tube'. He was particularly interested in the cathode rays produced by the procedure.

When Roentgen charged up a Crookes tube wrapped in cardboard, he was astonished to see something glowing on the other side of the room. It turned out to be a screen coated with a chemical called barium platinocyanide. According to existing theory, there was no way that the glow could be caused by cathode rays. Roentgen set about establishing what was going on. At one point, he glanced at the screen while he was holding an object between it and the Crookes tube. To his disbelief, he saw the bones of his hand unmistakably displayed. Roentgen initially doubted his sanity, writing to a friend that: 'I have discovered something interesting, but do not know whether or not my observations are correct.' But after seven weeks of painstaking investigation, he was convinced that his eyes had not deceived him. He gave a preliminary report to the Wurzburg Physical-Medical Society on 28 December, accompanied by an image of his wife's hand.

Wilhelm Roentgen, the German physicist who discovered X-rays.

By January, the world had gone X-ray mad. Roentgen was acclaimed as a medical genius, and in 1901 he was awarded the inaugural Nobel Prize for physics. Despite the justifiable fuss, Roentgen refused to patent his discovery, considering it a gift to mankind.

Y

YO-YO

The yo-yo is an ancient toy. It is the second oldest toy in the world after the doll, and a toy with a past that is tainted by disputes over ownership rights, allegations of a violent history, and has experienced a temperamental relationship with fashion.

The earliest record of yo-yos goes back some 2,500 years to Greece. Archaeologists have found vases depicting yo-yo use from this time, as well as rudimentary yo-yos made of terracotta, wood and metal. However, the yo-yo is generally agreed to have originated in China, and its popularity probably spread through trade routes. Throughout the 1800s, the yo-yo enjoyed popularity in England (where it was known as a 'bandalore' or 'quiz') and also especially among the aristocracy in France during the Revolution (where it was named *incroyable* or *l'emigrette*), where it was, allegedly, used as a device for stress relief. In 1789, the French Prince Louis XVII appeared in a portrait with his yo-yo.

The name 'yo-yo' probably came from the Philippines, although there is some confusion about its exact source. Some claim 'yo-yo' is a word native to the Philippines, meaning 'come come', or 'come back', although it is not used in this capacity today. The name was introduced to North America in the 1920s by a man called Pedro Flores, a Filipino immigrant. Working at a Santa Monica hotel, he would pass the time perfecting his yo-yo techniques – a popular hobby in the Philippines – and found his activities capturing the attention and imagination of passers-by. He started to make and sell yo-yos. To this day, there remains some controversy over the use of the yo-yo in the Philippines. Many believe that for a time it was used primarily as a weapon. Considerably larger than the yo-yo we recognise, 'with sharp edges and studs and attached to thick 20-foot ropes for flinging at enemies or prey', it must have been a terrifying and formidable missile if wielded with skill; it was certainly not a toy. While there is some evidence of such practices occurring, they are firmly denied by 'Duncan Yo-Yo', one of the main manufacturers of the toy today. Donald Duncan (1892–1971) bought Pedro Flores's yo-yo empire in 1929, by which time Flores had yo-yo factories in Santa Barbara, Los Angeles and Hollywood, producing some 300,000 yo-yos a day. Upon acquiring the company, Duncan registered as a trademark the name 'yo-yo', owning the rights to it until 1965 when, following an arduous lawsuit, it was declared a generic term. The yo-yos manufactured under Duncan displayed perhaps the only significant change in yo-yo design: a 'slip string' – a loop rather than a knot securing the yo-yo to its lead – which enabled users to make the toy 'sleep', and a host of tricks never considered before. Again, there are disagreements about who is responsible for this revolutionary modification. Some insist Donald Duncan himself took the yo-yo to this new, unanticipated plateau, while others maintain that the change occurred much earlier, back in the Philippines. Not to deprive Duncan of his dues entirely, he did pilot the alternative 'butterfly' design, a marginally more nimble model of yo-yo.

Sales slumped during World War II, but were revived in the 1960s when America worked itself into a veritable yo-yo frenzy. The effects of a yo-yo comedown and the financially ruinous court case of 1965 (resulting in bankruptcy for both parties) were balanced in the 1970s by the launch of ball-bearing yo-yos by Duncan (now owned by the Flambeau Products Corporation). First considered a bit of a gimmick by the hardcore yo-yoing community, the new possibilities presented by the design went someway to reviving interest. The last significant modification occurred in 2001 with the launch of the 'Freehand Yo-Yo', which swiftly became *the* yo-yo to own among professional, competitive yo-yoers.

Some popular yo-yo tricks include:

- Walk the Dog
- Man on the Flying Trapeze
- Elevator
- Lariat
- Around the World
- World Tour
- Over the Falls
- Three-Leaf Clover
- Tidal Wave/Skin the Cat
- Quickdraw/Gunslinger

#Z

ZIPPER

US Patent 1,219,881 20 March 1917

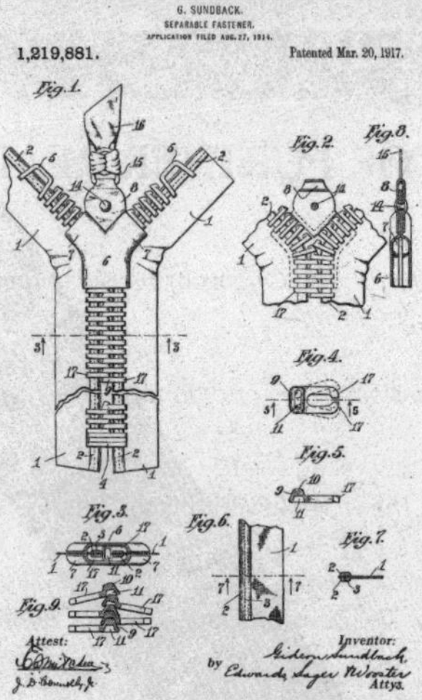

Sundback's zipper design.

The world's first zipper wasn't called a zipper and it didn't work very well, but the idea revolutionised the clothing industry. On 29 August 1893, a Chicago-based engineer named Whitcomb Judson (1836-1909) received a patent for a 'clasp-locker'. The mechanism was designed to replace the button-hooked laces then used to close the sides of high boots. It looked like a medieval torture instrument but it worked. To market the product, Judson teamed up with a businessman called Colonel Lewis Walker to found the Universal Fastener Company. Then the pair headed for the 1893 World's Fair in Chicago. They sold a grand total of 20 units, not to a boot manufacturer but to the US Postal Service. They were used to seal mailbags. In 1913, the design was improved by Gideon Sundback (1880-1954), a Swedish-born employee of the Universal Fastener Company. The modified fastener found a ready market in the form of the US military. The term 'zipper' was coined by the BF Goodrich Company in 1923 as an onomatopoeic descriptor for a successful line of rubber galoshes.

FOR FURTHER READING

Really Useful: The Origins of Everyday Things – Joel Levy

The Extraordinary Origins of Everyday Things – Charles Panati

The First of Everything – Dennis Sanders

The Book of Firsts – Patrick Robertson

Science & Technology Firsts – Leonard C Bruno

Eureka! An Illustrated History of Invention from the Wheel to the Computer – Edward DeBono

Famous First Facts – Joseph Nathan Kane

The Ideas Companion – Johnny Acton

A History of Everyday Things in England – Marjorie and CHB Quennell